STUDENT WORKBOOK

VOLUME 1

COLLEGE PHYSICS
A STRATEGIC APPROACH

KNIGHT • JONES • FIELD

RANDALL D. KNIGHT
CALIFORNIA POLYTECHNIC STATE UNIVERSITY,
SAN LUIS OBISPO

JAMES H. ANDREWS
YOUNGSTOWN STATE UNIVERSITY

PEARSON

Addison
Wesley

San Francisco   Boston   New York
Capetown   Hong Kong   London   Madrid   Mexico City
Montreal   Munich   Paris   Singapore   Sydney   Tokyo   Toronto

Editorial Director: Adam Black
Sponsoring Editor: Alice Houston
Development Manager: Michael Gillespie
Project Editor: Martha Steele
Managing Editor: Corinne Benson
Production Supervisor: Shannon Tozier
Production Service and Compositor: WestWords, Inc.
Illustrations: Precision Graphics
Text Design: Seventeenth Street Studios and WestWords, Inc.
Cover Design: Yvo Riezebos and Seventeenth Street Studios
Manufacturing Manager: Pam Augspurger
Text and Cover Printer: Bind-Rite Graphics
Cover Image: Ken Wilson; Papilio/CORBIS

The figures on page 8-2 of Chapter 8 are adapted from "Free-Body Diagrams Revisited — II" by James E. Court, *The Physics Teacher,* Volume 37, November 1999. © 1999 American Association of Physics Teachers. Reprinted by permission.

Many of the designations used by manufacturers and sellers to distinguish their products are claimed as trademarks. Where those designations appear in this book, and the publisher was aware of a trademark claim, the designations have been printed in initial caps or all caps.

ISBN: 0-8053-8209-7 Volume 1
ISBN: 0-8053-0626-9 Volume 2

Table of Contents

Dynamics Worksheets
Momentum Worksheets
Energy Worksheets

Preface

It is highly unlikely that one could learn to play the piano by only reading about it. Similarly, reading physics from a textbook is not the same as doing physics. To develop your ability to do physics, your instructor will assign problems to be solved both for homework and on tests. Unfortunately, it is our experience that jumping right into problem solving after reading and hearing about physics often leads to poor "playing" techniques and an inability to solve problems for which the student has not already been shown the solution (which isn't really "solving" a problem, is it?). Because improving your ability to solve physics problems is one of the major goals of your course, time spent developing techniques that will help you do this is well spent.

Learning physics, as in learning any skill, requires regular practice of the basic techniques. That is what this *Student Workbook* is all about. The workbook consists of exercises that give you an opportunity to practice techniques and strengthen your understanding of concepts presented in the textbook and in class. These exercises are intended to be done on a daily basis, right after the topics have been discussed in class and are still fresh in your mind. Successful completion of the workbook exercises will prepare you to tackle the more quantitative end-of-chapter homework problems in the textbook.

You will find that many of the exercises are *qualitative* rather than *quantitative*. They ask you to draw pictures, interpret graphs, use ratios, write short explanations, or provide other answers that do not involve calculations. A few math-skills exercises will ask you to explore the mathematical relationships and symbols used to quantify physics concepts, but do not require a calculator. The purpose of all of these exercises is to help you develop the basic thinking tools you'll later need for quantitative problem solving. It is highly recommended that you do these exercises *before* starting the end-of-chapter problems

One example from Chapter 4 illustrates the purpose of this *Student Workbook*. In that chapter, you will read about a technique called a "free-body diagram" that is helpful for solving problems involving forces. Sometimes students mistakenly think that the diagrams are used by the instructor only for teaching purposes and may be abandoned once Newton's laws are fully understood. On the contrary, professional physicists with decades of problem-solving experience still routinely use these diagrams to clarify the problem and set up the solution. Many of the other techniques practiced in the workbook, such as ray diagrams, graphing relationships, sketching field lines and equipotentials, etc., fall in the same category. They are used at all levels of physics, not just as a beginning exercise. And many of these techniques, such as analyzing graphs and exploring multiple representations of a situation, have important uses outside of physics. Time spent practicing these techniques will serve you well in other endeavors.

You will find that the exercises in this workbook are keyed to specific sections of the textbook in order to let you practice the new ideas introduced in that section. You should keep the text beside you as you work and refer to it often. You will usually find Tactics Boxes, figures, or examples in the textbook that are directly relevant to the exercises. When asked to draw figures or diagrams, you should attempt to draw them so that they look much like the figures and diagrams in the textbook.

Because the exercises go with specific sections in the text, you should answer them on the basis of information presented in *just* that section (and prior sections). You may have learned new ideas in Section 7 of a chapter, but you should not use those ideas when answering questions from Section 4. There will be ample opportunity in the Section 7 exercises to use that information there.

You will need a few "tools" to complete the exercises. Many of the exercises will ask you to *color code* your answers by drawing some items in black, others in red, and perhaps yet others in blue. You need to purchase a few colored pencils to do this. The authors highly recommend that you work in pencil, rather than ink, so that you can easily erase. Few are the individuals who make so few mistakes as to be able to work in ink! In addition, you'll find that a small, easily carried six-inch ruler will come in handy for drawings and graphs.

As you work your way through the textbook and this workbook, you will find that physics is a way of *thinking* about how the world works and why things happen as they do. We will primarily be interested in finding relationships, seeking explanations, and developing techniques to make use of these relationships, only secondarily in computing numerical answers. In many ways, the thinking tools developed in this workbook are what the course is all about. If you take the time to do these exercises regularly and to review the answers, in whatever form your instructor provides them, you will be well on your way to success in physics.

To the instructor: The exercises in this workbook can be used in many ways. You can have students work on some of the exercises in class as part of an active-learning strategy. Or you can do the same in recitation sections or laboratories. This approach allows you to discuss the answers immediately, to answer student questions, and to improvise follow-up exercises when needed. Having the students work in small groups (2 to 4 students) is highly recommended.

Alternatively, the exercises can be assigned as homework. The pages are perforated for easy tear-out, and the page breaks are in logical places so that you can assign the sections of a chapter that you would likely cover in one day of class. Exercises should be assigned immediately after presenting the relevant information in class and should be due at the beginning of the next class. Collecting them at the beginning of class, then going over two or three that are likely to cause difficulty, is an effective means of quickly reviewing major concepts from the previous class and launching a new discussion.

If used as homework, it is *essential* for students to receive *prompt* feedback. Ideally this would occur by having the exercises graded, with written comments, and returned at the next class meeting. Posting fairly detailed answers on a course website also works. Lack of prompt feedback can negate much of the value of these exercises. Placing similar qualitative/graphical questions on quizzes and exams, and telling students at the beginning of the term that you will do so, encourages students to take the exercises seriously and to check the answers.

One of the authors has been successful with assigning *all* exercises in the workbook as homework, collecting and grading them every day through Chapter 4, then collecting and grading them on about one-third of subsequent days on a random basis. The other author uses the exercises in class as immediate practice of the techniques demonstrated in the text and on the chalkboard. Student feedback from end-of-term questionnaires reveals three prevalent attitudes toward the workbook exercises:

  i. They think it is an unreasonable amount of work.
  ii. They agree that the assignments force them to keep up and not get behind.
  iii. They recognize, by the end of the term, that the workbook is a valuable learning tool.

However you choose to use these exercises, they will significantly strengthen your students' conceptual understanding of physics.

Following the workbook exercises are optional Dynamics Worksheets, Momentum Worksheets, and Energy Worksheets for use with end-of-chapter problems in Parts I and II of the textbook. Their use is recommended to help students acquire good problem-solving habits early in the course. End-of-chapter problems marked with the 🖉 icon are intended to be done on worksheets.

Answers to all workbook exercises are provided as pdf files on the *Media Manager* CD-ROM.

Acknowledgments: The authors would like to thank Rebecca L. Sobinovsky and Jared Sterzer.

1 Concepts of Motion and Mathematical Background

1.1 Motion: A First Look

Exercises 1–5: Draw a motion diagram for each motion described below.
- Use the particle model to represent the object as a particle.
- Six to eight dots are appropriate for most motion diagrams.
- Number the positions in order, as shown in Figure 1.4 in the text.
- Be neat and accurate!

1. A car accelerates forward from a stop sign. It eventually reaches a steady speed of 45 mph.

2. An elevator starts from rest at the 100th floor of the Empire State Building and descends, with no stops, until coming to rest on the ground floor. (Draw this one *vertically* since the motion is vertical.)

3. A skier starts *from rest* at the top of a 30° snow-covered slope and steadily speeds up as she skies to the bottom. (Orient your diagram as seen from the *side*. Label the 30° angle.)

4. The space shuttle orbits the earth in a circular orbit, completing one revolution in 90 minutes.

5. Bob throws a ball at an upward 45° angle from a third-story balcony. The ball lands on the ground below.

Exercises 6–9: For each motion diagram, write a short description of the motion of an object that will match the diagram. Your descriptions should name *specific* objects and be phrased similarly to the descriptions of Exercises 1 to 5. Note the axis labels on Exercises 8 and 9.

6.

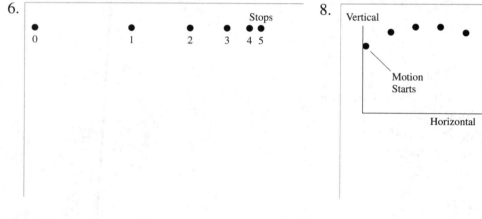

7.

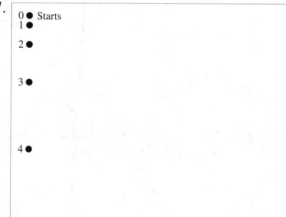

8.

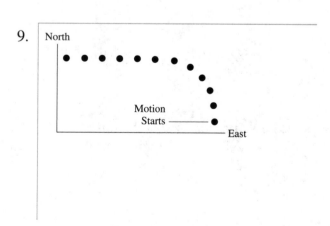

9.

1.2 Position and Time: Putting Numbers on Nature

10. Redraw each of the motion diagrams from Exercises 1 to 3 in the space below. Add a coordinate axis to each drawing and label the initial and final positions. Draw an arrow on your diagram to represent the displacement from the beginning to the end of the motion.

11. In the picture below Joe starts walking casually at constant speed from his house on Main Street to the bus stop 200 m down the street. When he is halfway there, he sees the bus and steadily speeds up until he reaches the bus stop.

    a. Draw a motion diagram in the street of the picture to represent Joe's motion.

    b. Add a coordinate axis below your diagram with Joe's house as the origin. Label Joe's initial position at the start of his walk as x_1, his position when he sees the bus as x_2, and his final position when he arrives at the bus stop as x_3. Draw arrows above the coordinate axis to represent Joe's displacement from his initial position to his position when he first sees the bus and the displacement from where he sees the bus to the bus stop. Label these displacements Δx_1 and Δx_2, respectively.

    c. Repeat part b in the space below but with the origin at the location where Joe starts to speed up.

    d. How do the displacement arrows change when you change the location of the origin?

1.3 Velocity

12. A moth flies a distance of 3 m in only one-third of a second.
    a. What does the ratio 3/(1/3) tell you about the moth's motion? Explain.

    b. What does the ratio (1/3)/3 tell you about the moth's motion?

    c. How far would the moth fly in one tenth of a second?

    d. How long does it take the moth to fly 4 m?

13. a. If someone drives at 25 miles per hour, is it necessary that they do so for an hour?

    b. Is it necessary to have a cubic centimeter of gold to say that gold has a density of 19.3 grams per cubic centimeter? Explain.

1.4 A Sense of Scale: Significant Figures, Scientific Notation, and Units

14. How many significant figures does each of the following numbers have?

    a. 6.21 _____   e. 0.0621 _____   i. 1.0621 _____

    b. 62.1 _____   f. 0.620 _____   j. 6.21×10^3 _____

    c. 6210 _____   g. 0.62 _____   k. 6.21×10^{-3} _____

    d. 6210.0 _____   h. .62 _____   l. 62.1×10^3 _____

15. Compute the following numbers, applying the significant figure standards adopted for this text.

    a. $33.3 \times 25.4 =$ _____   e. $2.345 \times 3.321 =$ _____

    b. $33.3 - 25.4 =$ _____   f. $(4.32 \times 1.23) - 5.1 =$ _____

    c. $33.3 \div 45.1 =$ _____   g. $33.3^2 =$ _____

    d. $33.3 \times 45.1 =$ _____   h. $\sqrt{33.3} =$ _____

16. Express the following numbers and computed results in scientific notation, paying attention to significant figures.

    a. $9{,}827 =$ _____   d. $32{,}014 \times 47 =$ _____

    b. $0.000000550 =$ _____   e. $0.059 \div 2{,}304 =$ _____

    c. $3{,}200{,}000 =$ _____   f. $320. \times 0.050 =$ _____

17. Convert the following to SI units. Work across the line and show all steps in the conversion. Use scientific notation and apply the proper use of significant figures. **Note:** Think carefully about g and h. A picture may help.

    a. $9.12 \ \mu s \times$

    b. $3.42 \ \text{km} \times$

    c. $44 \ \text{cm/ms} \times$

    d. $80 \ \text{km/hr} \times$

    e. $60 \ \text{mph} \times$

    f. $8 \ \text{in} \times$

    g. $14 \ \text{in}^2 \times$

    h. $250 \ \text{cm}^3 \times$

18. Use Tables 1.4 and 1.5 and Examples 1.2 and 1.3 to assess whether or not the following statements are *reasonable*.

    a. Joe is 180 cm tall.

    b. I rode my bike to campus at a speed of 50 m/s.

    c. A skier reaches the bottom of the hill going 25 m/s.

    d. I can throw a ball a distance of 2 km.

    e. I can throw a ball at a speed of 50 km/hr.

    f. Joan's newborn baby has a mass of 33 kg.

    g. A typical hummingbird has a mass of 3.3 g.

1.5 Vectors and Motion: A First Look

19. For the following motion diagrams, draw an arrow to indicate the displacement vector between the initial and final positions.

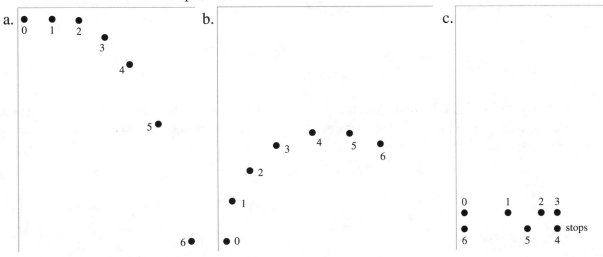

20. In Exercise 19, is the object's displacement equal to the distance the object travels? Explain.

21. Draw and label the vector sum $\vec{A} + \vec{B}$.

a. 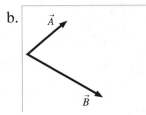   b.   c.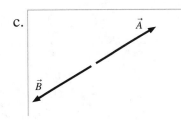

Exercises 22–26: Draw a motion diagram for each motion described below.
- Use the particle model.
- Show and label the *velocity* vectors.

22. Galileo drops a ball from the Leaning Tower of Pisa. Consider the ball's motion from the moment it leaves his hand until a microsecond before it hits the ground. Your diagram should be vertical.

23. A rocket-powered car on a test track accelerates from rest to a high speed, then coasts at constant speed after running out of fuel. Draw a dotted line across your diagram to indicate the point at which the car runs out of fuel.

24. A bowling ball being returned from the pin area to the bowler starts out rolling at a constant speed. It then goes up a ramp and exits onto a level section at very low speed. You'll need 10 or 12 points to indicate the motion clearly.

25. A car is parked on a hill. The brakes fail, and the car rolls down the hill with an ever-increasing speed. At the bottom of the hill it runs into a thick hedge and gently comes to a halt.

26. Andy is standing on the street. Bob is standing on the second-floor balcony of their apartment, about 30 feet back from the street. Andy throws a baseball to Bob. Consider the ball's motion from the moment it leaves Andy's hand until a microsecond before Bob catches it.

1.6 Making Models: The Power of Physics

27. One of the difficulties some students have in beginning physics is with the use of algebra involving unfamiliar symbols. As a warmup for what's to come, algebraically solve the following equations for the specified variables in terms of the other symbols given:

    a. Solve for t: $\quad v = v_0 + at$

    b. Solve for x: $\quad t = \left(\dfrac{2x}{a}\right)^{1/2}$

    c. Solve for s: $\quad \dfrac{1}{s} + \dfrac{1}{s'} = \dfrac{1}{f}$

    d. Eliminate T from the two equations and solve for a: $\quad$ $T - m_1 g = m_1 a$
    $\quad m_2 g - T = m_2 a$

28. On the axes below, sketch a graph of y versus x if y is given by the equation shown. Assume that m and b are both positive numbers. The goal is to sketch a graph with the proper *shape*.

    a. $y = mx + b$

    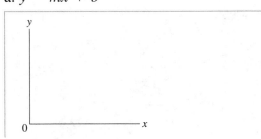

    b. $y = mx^2$

    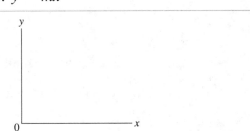

    c. $y = m/x^2$

    d. $y^2 = mx$

1.7 Where Do We Go from Here?

No exercises for this section.

2 Motion in One Dimension

2.1 Describing Motion

1. Sketch position-versus-time graphs for the following motions. Include a numerical scale on both axes with units that are *reasonable* for this motion. Some numerical information is given in the problem, but for other quantities make reasonable estimates.

   Note: A *sketched* graph simply means hand-drawn, rather than carefully measured and laid out with a ruler. But a sketch should still be neat and as accurate as is feasible by hand. It also should include labeled axes and, if appropriate, tick-marks and numerical scales along the axes.

   a. A student walks to the bus stop, waits for the bus, then rides to campus. Assume that all the motion is along a straight street.

   b. A student walks slowly to the bus stop, realizes he forgot his paper that is due, and *quickly* walks home to get it.

   c. The quarterback drops back 10 yards from the line of scrimmage, then throws a pass 20 yards to the tight end, who catches it and sprints 20 yards to the goal. Draw your graph for the *football*. Think carefully about what the slopes of the lines should be.

2. The position-versus-time graph below shows the position of an object moving in a straight line for 12 seconds.

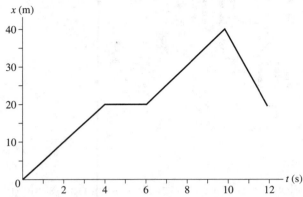

a. What is the position of the object at 2 s, 6 s, and 10 s after the start of the motion?

   At 2 s: _____

   At 6 s: _____

   At 10 s: _____

b. What is the object's velocity during the first 4 s of motion?

c. What is the object's velocity during the interval from $t = 4$ s to $t = 6$ s?

d. What is the object's velocity during the four seconds from $t = 6$ s to $t = 10$ s?

e. What is the object's velocity during the final two seconds from $t = 10$ s to $t = 12$ s?

f. Draw a motion diagram below to represent the entire 12 s of motion.

3. Interpret the following position-versus-time graphs by writing a very short "story" of what is happening. Be creative! Have characters and situations! Simply saying that "a car moves 100 meters to the right" doesn't qualify as a story. Your stories should make *specific reference* to information you obtain from the graphs, such as distances moved or time elapsed.

a. Moving car

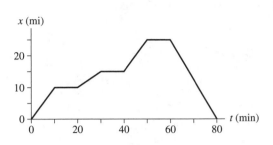

b. Sprinter

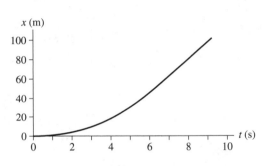

c. Two football players

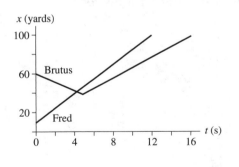

4. Can you give an interpretation to this position-versus-time graph? If so, then do so. If not, why not?

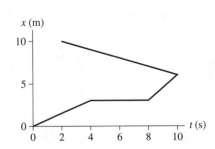

2.2 Uniform Motion

5. Sketch position-versus-time graphs for the following motions. Include appropriate numerical scales along both axes. A small amount of computation may be necessary.

   a. A parachutist opens her parachute at an altitude of 1500 m. She then descends slowly to earth at a steady speed of 5 m/s. Start your graph as her parachute opens.

   b. Trucker Bob starts the day 120 miles west of Denver. He drives east for 3 hours at a steady 60 miles/hour before stopping for his coffee break. Let Denver be located at $x = 0$ mi and assume that the x-axis points to the east.

   c. Quarterback Bill throws the ball to the right at a speed of 15 m/s. It is intercepted 45 m away by Carlos, who is running to the left at 7.5 m/s. Carlos carries the ball 60 m to score. Let $x = 0$ m be the point where Bill throws the ball. Draw the graph for the *football*.

6. The figure shows a position-versus-time graph for the motion of objects A and B that are moving along the same axis.

   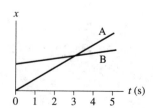

   a. At the instant $t = 1$ s, is the speed of A greater than, less than, or equal to the speed of B? Explain.

   b. Do objects A and B ever have the *same* speed? If so, at what time or times? Explain.

7. Draw both a position-versus-time graph *and* a velocity-versus-time graph for an object that is at rest at $x = 1$ m.

8. The figure shows six frames from the motion diagram of two moving cars, A and B.

   a. Draw both a position-versus-time graph and a velocity-versus-time graph. Show the motion of *both* cars on each graph. Label them A and B.

   b. Do the two cars ever have the same position at one instant of time?

      If so, in which frame number (or numbers)? _____

      Draw a vertical line through your graphs of part a to indicate this instant of time.

9. Below are four position-versus-time graphs. For each, draw the corresponding velocity-versus-time graph directly below it. A vertical line drawn through both graphs should connect the velocity v_x at time t with the position s at the *same* time t. There are no numbers, but your graphs should correctly indicate the *relative* speeds.

   a. 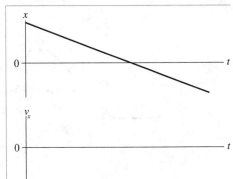   b.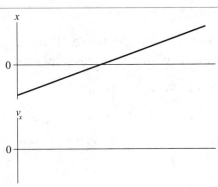

   c. 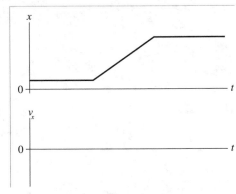   d.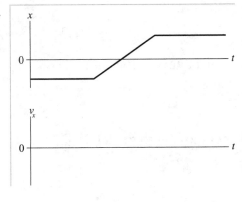

10. Below are two velocity-versus-time graphs. For each:
    - Draw the corresponding position-versus-time graph.
    - Give a written description of the motion.

    Assume that the motion takes place along a horizontal line and that $x_0 = 0$.

    a.
    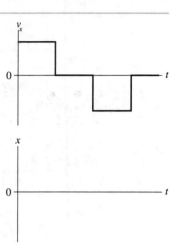

    b.
    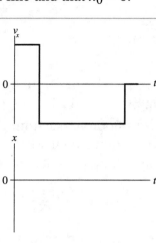

11. The figure shows a position-versus-time graph for a moving object. At which lettered point or points:

    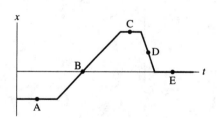

    a. Is the object moving the slowest? _____

    b. Is the object moving the fastest? _____

    c. Is the object at rest? _____

    d. Does the object have a constant nonzero velocity? _____

    e. Is the object moving to the left? _____

2.3 Motion with Changing Velocity

12. Below are two position-versus-time graphs. For each, draw the corresponding velocity-versus-time graph directly below it. A vertical line drawn through both graphs should connect the velocity v_x at time t with the position x at the *same* time t. There are no numbers, but your graphs should correctly indicate the *relative* speeds.

a.

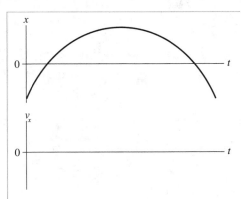

b.

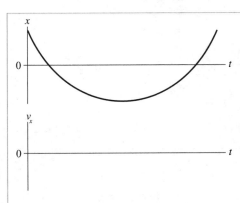

13. The figure shows six frames from the motion diagram of two moving cars, A and B.

    a. Draw both a position-versus-time graph and a velocity-versus-time graph. Show *both* cars on each graph. Label them A and B.

   b. Do the two cars ever have the same position at one instant of time?

      If so, in which frame number (or numbers)? _____

      Draw a vertical line through your graphs of part a to indicate this instant of time.

   c. Do the two cars ever have the same velocity at one instant of time?

      If so, between which two frames? _____

14. For each of the following motions, draw
    - A motion diagram,
    - A position-versus-time graph, and
    - A velocity-versus-time graph.

    a. A car starts from rest, steadily speeds up to 40 mph in 15 s, moves at a constant speed for 30 s, then comes to a halt in 5 s.

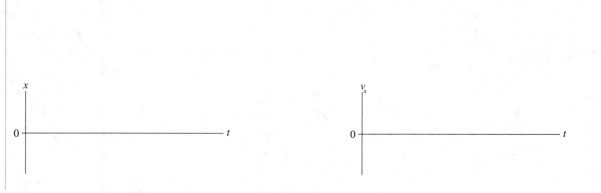

    b. A rock is dropped from a bridge and steadily speeds up as it falls. It is moving at 30 m/s when it hits the ground 3 s later. Think carefully about the signs.

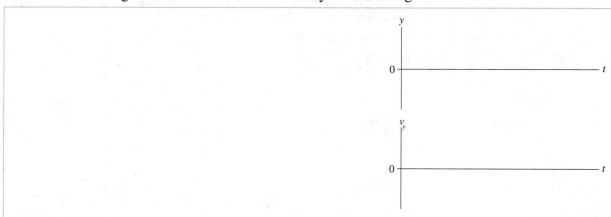

    c. A pitcher winds up and throws a baseball with a speed of 40 m/s. One-half second later the batter hits a line drive with a speed of 60 m/s. The ball is caught 1 s after it is hit. From where you are sitting, the batter is to the right of the pitcher. Draw your motion diagram and graph for the *horizontal* motion of the ball.

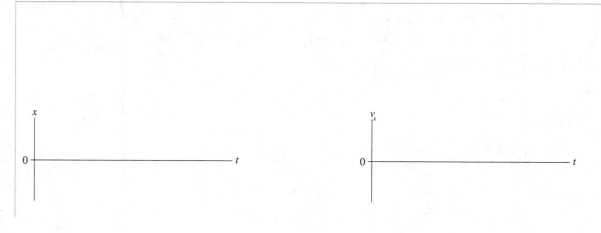

15. Below are shown four velocity-versus-time graphs. For each:
    - Draw the corresponding position-versus-time graph.
    - Give a written description of the motion.

    Assume that the motion takes place along a horizontal line and that $x_0 = 0$.

a.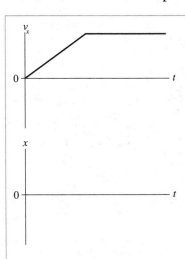

b.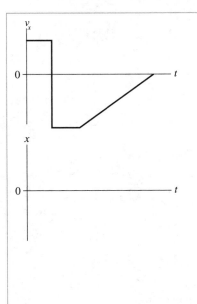

c.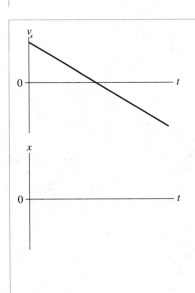

d.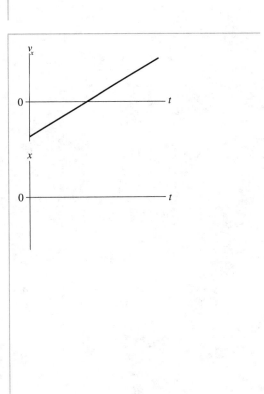

2.4 Acceleration

16. The four motion diagrams below show an initial point 0 and a final point 1. A pictorial representation would define the five symbols: x_0, x_1, v_{0x}, v_{1x}, and a_x for horizontal motion and equivalent symbols with y for vertical motion. Determine whether each of these quantities is positive, negative, or zero. Give your answer by writing +, −, or 0 in the table below.

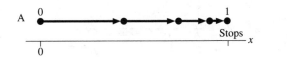

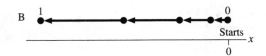

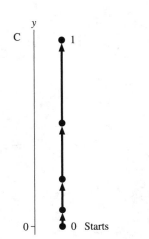

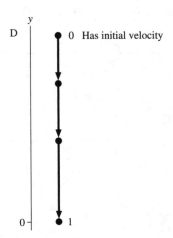

	A	B	C	D
x_0 or y_0				
x_1 or y_1				
v_{0x} or v_{0y}				
v_{1x} or v_{1x}				
a_x or a_y				

17. The three symbols x, v_x, and a_x have eight possible combinations of *signs*. For example, one combination is $(x, v_x, a_x) = (+, -, +)$.

   a. List all eight combinations of signs for x, v_x, a_x.

   1. _____   5. _____

   2. _____   6. _____

   3. _____   7. _____

   4. _____   8. _____

b. For each of the eight combinations of signs you identified in part a:
  • Draw a four-dot motion diagram of an object that has these signs for x, v_x, and a_x.
  • Draw the diagram *above* the axis whose number corresponds to part a.
  • Use **black** and **red** for your $\vec{v}$ and $\vec{a}$ vectors. Be sure to label the vectors.

1. ———————————————————————————— x
                              0

2. ———————————————————————————— x
                              0

3. ———————————————————————————— x
                              0

4. ———————————————————————————— x
                              0

5. ———————————————————————————— x
                              0

6. ———————————————————————————— x
                              0

7. ———————————————————————————— x
                              0

8. ———————————————————————————— x
                              0

2.5 Motion with Constant Acceleration

18. For each of the following situations, provide a description and a motion diagram.
    a. $a_x = 0$ but $v_x \neq 0$.

    b. $v_x = 0$ but $a_x \neq 0$.

    c. $v_x < 0$ and $a_x > 0$.

19. The quantity y is proportional to the square of x, and $y = 36$ when $x = 3$.
    a. Write an equation to represent this quadratic relationship for all y and x.

    b. Find y if $x = 5$. _____   c. Find x if $y = 16$. _____

    d. By what factor must x change for the value of y to double? _____

    e. Compare your equation in part a to the equation from your text relating Δx and Δt,

    $\Delta x = \dfrac{1}{2}a_x\Delta t^2$. Which quantity assumes the role of x? Which quantity assumes the role of y?

    What is the constant of proportionality relating Δx and Δt?

20. Below are three velocity-versus-time graphs. For each, draw the corresponding acceleration-versus-time graph and draw a motion diagram below the graphs.

a.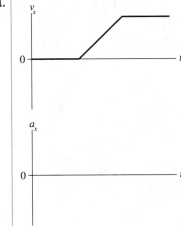

b.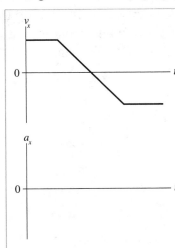

c.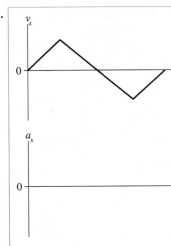

2.6 Solving One-Dimensional Motion Problems

21. Draw a pictorial representation of each situation described below. That is, (i) sketch the situation, showing appropriate points in the motion, (ii) establish a coordinate system on your sketch, and (iii) define appropriate symbols for the known and unknown quantities. **Do not solve.**

   a. A bicyclist starts from rest and accelerates at 4.0 m/s^2 for 3.0 s. The cyclist then travels for 20 s at a constant speed before slowing to a stop in just 2.0 s. How far does the cyclist travel?

   Known

   Find

   b. You are driving your car at 12 m/s when a deer jumps in front of your car. What is the shortest stopping distance for your car if your reaction time is 0.80 s and your car brakes at 6.0 m/s^2?

   Known

   Find

   c. At the snap, Quarterback Bill throws the football to the right at a speed of 15 m/s. It is intercepted 45 m away by Carlos, who accelerates at 2.5 m/s from rest to a speed of 7.5 m/s to the left and carries the ball for a total distance of 60 m to score. How much time has elapsed on the game clock?

   Known

   Find

2.7 Free Fall

22. A ball is thrown straight up into the air. At each of the following instants, is the ball's acceleration g, $-g$, 0, $< g$, or $> g$?

    a. Just after leaving your hand? ....

    b. At the very top (maximum height)? ....

    c. Just before hitting the ground? ....

23. A ball is thrown straight up into the air. It reaches height h, then falls back down to the ground. On the axes below, graph the ball's acceleration from an instant after it leaves the thrower's hand until an instant before it hits the ground. Indicate on your graph the times during which the ball is moving upwards, at its peak, and moving downwards.

24. On a single graph, using the axes below, graph

    a. the acceleration of a rock dropped from a bridge into the river below, and

    b. the acceleration of a rock thrown (not dropped) from a bridge into the river below.

    Be sure to label your graphs.

3 Vectors and Motion in Two Dimensions

3.1 Using Vectors

1. Use a figure and the properties of vector addition to show that vector addition is associative. That is, show that
$$(\vec{A} + \vec{B}) + \vec{C} = \vec{A} + (\vec{B} + \vec{C})$$

2. Draw and label the vector $2\vec{A}$ and the vector $\frac{1}{2}\vec{A}$.

Exercises 3–5: Draw and label the vector difference $\vec{A} - \vec{B}$.

3.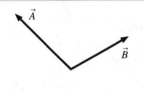

4.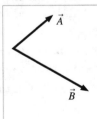

5.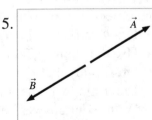

6. Given vectors $\vec{A}$ and $\vec{B}$ below, find the vector $\vec{C} = 2\vec{A} - 3\vec{B}$.

3.2 Using Vectors on Motion Diagrams

7. The figure below shows the positions of a moving object in three successive frames of film. Draw and label the velocity vector $\vec{v}_0$ for the motion from 0 to 1 and the vector $\vec{v}_1$ for the motion from 1 to 2. Then draw the vector $\vec{v}_1 - \vec{v}_0$ with its tail on point 1.

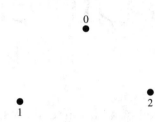

8. A car enters an icy intersection traveling at 16 m/s due north. After a violent collision with a truck, the car slides away moving 12 m/s due east. Draw arrows on the picture below to show (i) the car's velocity $\vec{v}_0$ when entering the intersection, (ii) its velocity $\vec{v}_1$ when leaving, and (iii) the car's change in velocity $\Delta\vec{v} = \vec{v}_1 - \vec{v}_0$ due to the collision.

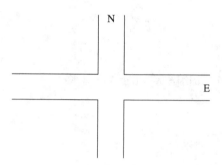

Exercises 9–10: The figures below show an object's position in three successive frames of film. The object is moving in the direction $0 \rightarrow 1 \rightarrow 2$. For each diagram:

- Draw and label the initial and final velocity vectors $\vec{v}_0$ and $\vec{v}_1$. Use **black**.
- Use the steps of Tactics Box 3.2 to find the change in velocity $\Delta\vec{v}$.
- Draw and label $\vec{a}$ at the proper location on the motion diagram. Use **red**.
- Determine whether the object is speeding up, slowing down, or moving at a constant speed. Write your answer beside the diagram.

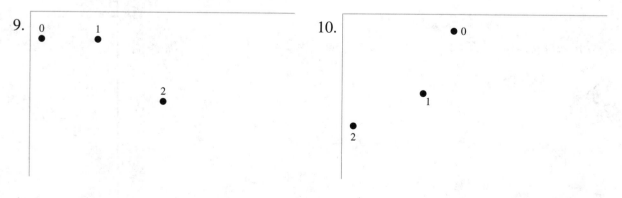

3.3 Coordinate Systems and Vector Components

3.4 Motion on a Ramp

Exercises 11–13: Draw and label the *x*- and *y*-component vectors of the vector shown.

11.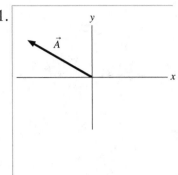

12.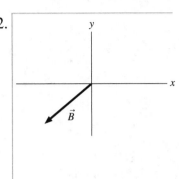

13.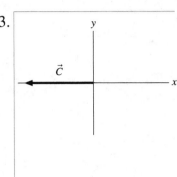

Exercises 14–16: Determine the numerical values of the *x*- and *y*-components of each vector.

14.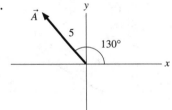

$A_x =$ _____

$A_y =$ _____

15.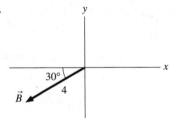

$B_x =$ _____

$B_y =$ _____

16.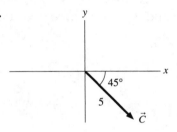

$C_x =$ _____

$C_y =$ _____

17. What is the vector sum $\vec{D} = \vec{A} + \vec{B} + \vec{C}$ of the three vectors defined in Exercises 14–16?

$D_x =$ _____ $D_y =$ _____

18. Can a vector have a component equal to zero and still have nonzero magnitude? Explain.

19. Can a vector have zero magnitude if one of its components is nonzero? Explain.

Exercises 20–22: For each vector:
- Draw the vector on the axes provided.
- Draw and label an angle θ to describe the direction of the vector.
- Find the magnitude and the angle of the vector.

20. $A_x = 3$, $A_y = -2$

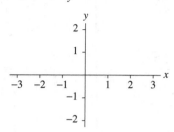

$A =$ _____

$\theta =$ _____

21. $B_x = -2$, $B_y = 2$

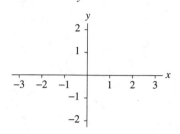

$B =$ _____

$\theta =$ _____

22. $C_x = 0$, $C_y = -2$

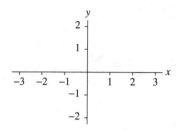

$C =$ _____

$\theta =$ _____

Exercises 23–25: Define vector $\vec{A} = (5, 30°$ above the horizontal). Determine the components A_x and A_y in the three coordinate systems shown below. Show your work below the figure.

23.

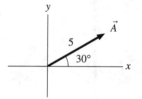

$A_x =$ _____

$A_y =$ _____

24.

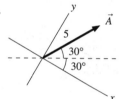

$A_x =$ _____

$A_y =$ _____

25.

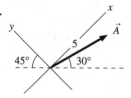

$A_x =$ _____

$A_y =$ _____

26. The figure shows a ramp and a ball that rolls along the ramp. Draw vector arrows on the figure to show the ball's acceleration at each of the lettered points A to E (or write $\vec{a} = \vec{0}$, if appropriate).

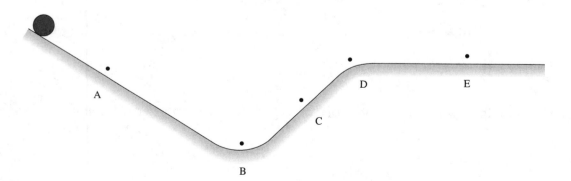

3.5 Relative Motion

27. On a ferry moving steadily forward in still water at 5 m/s, a passenger walks towards the back of the boat at 2 m/s. (i) Write a symbolic equation to find the velocity of the passenger with respect to the water using $(v_x)_{AB}$ notation. (ii) Substitute the appropriate values into your equation to determine the value of that velocity.

28. A boat crossing a river can move at 5 m/s with respect to the water. The river is flowing to the right at 3 m/s. In (a), the boat points straight across the river and is carried downstream by the water. In (b), the boat is angled upstream by the amount needed for it to travel straight across the river. For each situation, draw the velocity vectors $\vec{v}_{RS}$ of the river with respect to the shore, $\vec{v}_{BS}$ of the boat with respect to the river, and $\vec{v}_{BS}$ of the boat with respect to the shore,

a.

Start

b.

Finish

Start

29. Ryan, Samantha, and Tomas are driving their convertibles. At the same instant, they each see a jet plane with an instantaneous velocity of 200 m/s and an acceleration of 5 m/s^2.

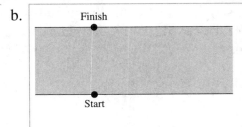

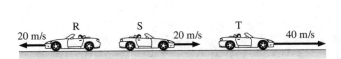

   a. Rank in order, from largest to smallest, the jet's *speed* v_R, v_S, and v_T according to Ryan, Samantha, and Tomas. Explain.

   b. Rank in order, from largest to smallest, the jet's *acceleration* a_R, a_S, and a_T according to Ryan, Samantha, and Tomas. Explain.

30. An electromagnet on the ceiling of an airplane holds a steel ball. When a button is pushed, the magnet releases the ball. The experiment is first done while the plane is parked on the ground, and the point where the ball hits the floor is marked with an X. Then the experiment is repeated while the plane is flying level at a steady 500 mph. Does the ball land slightly in front of the X (toward the nose of the plane), on the X, or slightly behind the X (toward the tail of the plane)? Explain.

31. Zack is driving past his house. He wants to toss his physics book out the window and have it land in his driveway. If he lets go of the book exactly as he passes the end of the driveway, should he direct his throw outward and toward the front of the car (throw 1), straight outward (throw 2), or outward and toward the back of the car (throw 3)? Explain.

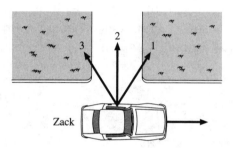

32. Yvette and Zack are driving down the freeway side by side with their windows rolled down. Zack wants to toss his physics book out the window and have it land in Yvette's front seat. Should he direct his throw outward and toward the front of the car (throw 1), straight outward (throw 2), or outward and toward the back of the car (throw 3)? Explain.

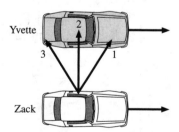

3.6 Motion in Two Dimensions: Projectile Motion

33. Complete the motion diagram for this trajectory, showing velocity and acceleration vectors.

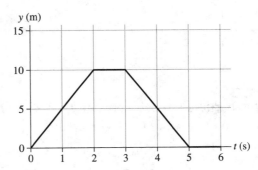

34. A particle moving along a trajectory in the *xy*-plane has the *x*-versus-*t* graph and the *y*-versus-*t* graph shown below.

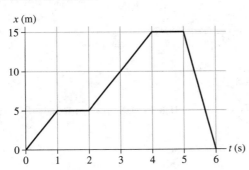

a. Use the grid below to draw a *y*-versus-*x* graph of the trajectory.

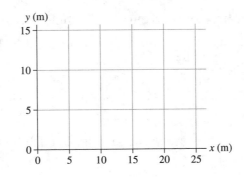

b. Draw the particle's velocity vector at $t = 3.5$ s on your graph.

35. The trajectory of a particle is shown below. The particle's position is indicated with dots at 1-second intervals. The particle moves between each pair of dots at constant speed.

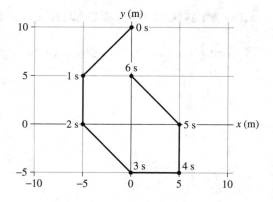

a. Draw x-versus-t and y-versus-t graphs for the particle.

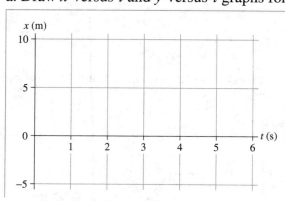
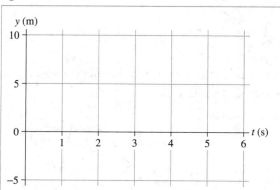

b. Is the particle's speed between $t = 5$ s and $t = 6$ s greater than, less than, or equal to its speed between $t = 1$ s and $t = 2$ s? Explain.

36. A projectile is launched over horizontal ground at an angle between 0° and 90°.
   a. Is there any point on the trajectory where $\vec{v}$ and $\vec{a}$ are parallel to each other? If so, where?

   b. Is there any point where $\vec{v}$ and $\vec{a}$ are perpendicular to each other? If so, where?

   c. Which of the following remain constant throughout the entire trajectory: x, y, v, v_x, v_y, a_x, a_y?

3.7 Projectile Motion: Solving Problems

37. The figure shows a ball that rolls down a quarter-circle ramp, then off a cliff. Sketch the ball's trajectory from the instant it is released until it hits the ground.

38. a. A cart that is rolling at constant velocity fires a ball straight up. When the ball comes back down, will it land in front of the launching tube, behind the launching tube, or directly in the tube? Explain.

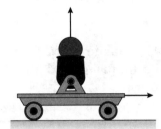

   b. Will your answer change if the cart is accelerating in the forward direction? If so, how?

39. A rock is thrown from a bridge at an angle 30° below horizontal.

   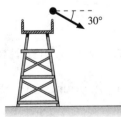

   a. Sketch the rock's trajectory on the figure.

   b. Immediately after the rock is released, is the magnitude of its acceleration greater than, less than, or equal to *g*? Explain.

   c. At the instant of impact, is the rock's speed greater than, less than, or equal to the speed with which it was thrown? Explain.

40. Four balls are simultaneously launched with the same speed from the same height h above the ground. At the same instant, ball 5 is released from rest at the same height. Rank in order, from shortest to longest, the amount of time it takes each of these balls to hit the ground. (Some may be simultaneous.)

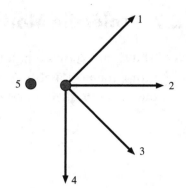

Order:

Explanation:

41. Rank in order, from shortest to longest, the amount of time it takes each of these projectiles to hit the ground. (Some may be simultaneous.)

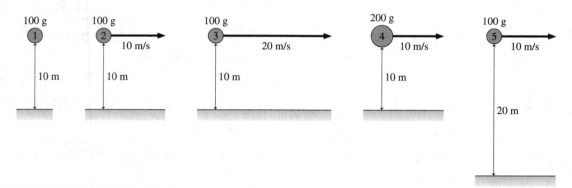

Order:

Explanation:

3.8 Motion in Two Dimensions: Circular Motion

42. a. The crankshaft in your car rotates at 3000 rpm. What is the frequency in revolutions per second?

  b. A record turntable rotates at 33.3 rpm. What is the period in seconds?

43. The dots of a motion diagram are shown below for an object in uniform circular motion. Carefully complete the diagram.
    - Draw and label the velocity vectors $\vec{v}$. Use a **black** pen or pencil.
    - Draw and label the acceleration vectors $\vec{a}$. Use a **red** pen or pencil.

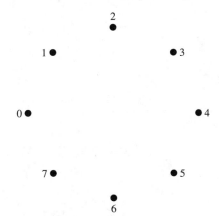

44. The figure shows three points on a steadily rotating wheel.
    a. Draw the velocity vectors at each of the three points.
    b. Rank in order, from largest to smallest, the speeds v_1, v_2, and v_3 of these points.

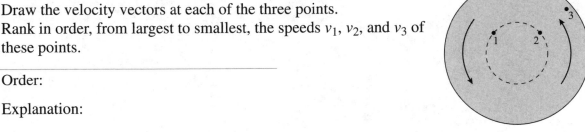

    Order:

    Explanation:

45. An object is traveling in a circle of radius r at constant speed v.
    a. By what factor does the object's acceleration change if its speed is doubled and the radius is unchanged? _____

    b. By what factor does the acceleration change if the radius of the circle is doubled and its speed is unchanged? _____

    c. By what factor does the acceleration change if the period of the motion is doubled without changing the size of the circle? _____

4 Forces and Newton's Laws of Motion

4.1 What Causes Motion?

4.2 Force

1. Using the particle model, represent the force a person exerts on a table when (a) pulling it to the right across a level floor with a force of magnitude F, (b) pulling it to the left across a level floor with force $2F$, and (c) *pushing* it to the right across a level floor with force F.

   a. Table pulled right
      with force F

   b. Table pulled left
      with force $2F$

   c. Table pushed right
      with force F

2. Two or more forces are shown on the objects below. Draw and label the net force $\vec{F}_{net}$.

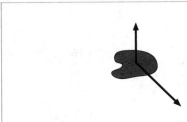

3. Two or more forces are shown on the objects below. Draw and label the net force $\vec{F}_{net}$.

4.3 A Short Catalog of Forces

4.4 Identifying Forces

Exercises 4–8: Follow the six-step procedure of Tactics Box 4.2 to identify and name all the forces acting on the object.

4. An elevator suspended by a cable is descending at constant velocity.

5. A compressed spring is pushing a block across a rough horizontal table.

6 A brick is falling from the roof of a three-story building.

7. Blocks A and B are connected by a string passing over a pulley Block B is falling and dragging block A across a frictionless table. Let block A be "the system" for analysis.

8. A rocket is launched at a 30° angle. Air resistance is not negligible.

4.5 What Do Forces Do?

9. The figure shows an acceleration-versus-force graph for an object of mass m. Data have been plotted as individual points, and a line has been drawn through the points.

   Draw and label, directly on the figure, the acceleration-versus-force graphs for objects of mass

   a. $2m$ b. $0.5m$

   Use triangles ▲ to show four points for the object of mass $2m$, then draw a line through the points. Use squares ■ for the object of mass $0.5m$.

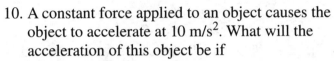

Acceleration vs. Force (rubber bands)

10. A constant force applied to an object causes the object to accelerate at 10 m/s². What will the acceleration of this object be if

    a. The force is doubled? _____   b. The mass is doubled? _____

    c. The force is doubled *and* the mass is doubled? _____

    d. The force is doubled *and* the mass is halved? _____

11. A constant force applied to an object causes the object to accelerate at 8 m/s². What will the acceleration of this object be if

    a. The force is halved? _____   b. The mass is halved? _____

    c. The force is halved *and* the mass is halved? _____

    d. The force is halved *and* the mass is doubled? _____

12. The quantity y is inversely proportional to x and $y = 4$ when $x = 9$.
    a. Write an equation to represent this inverse relationship for all y and x.

    b. Find y if $x = 12$ _____   c. Find x if $y = 36$ _____

    d. Compare your equation in part a to the equation from your text relating a and m, $a = \dfrac{F}{m}$.

       Which quantity assumes the role of x? _____
       Which quantity assumes the role of y? _____
       What is the constant of proportionality relating a and m?

4.6 Newton's Second Law

13. Forces are shown on three objects. For each:
    a. Draw and label the net force vector. Do this right on the figure.
    b. Below the figure, draw and label the object's acceleration vector.

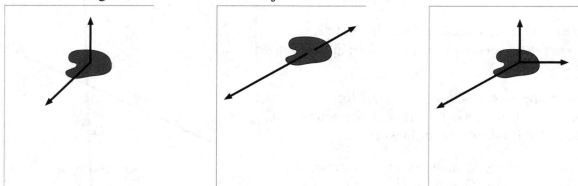

14. In the figures below, one force is missing. Use the given direction of acceleration to determine the missing force and draw it on the object. Do all work directly on the figure.

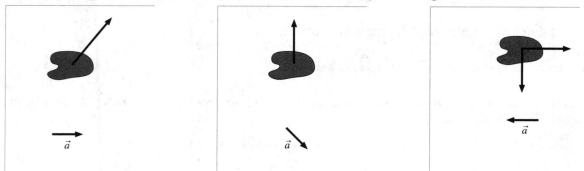

15. Below are two motion diagrams for a particle. Draw and label the net force vector at point 2.

16. Below are two motion diagrams for a particle. Draw and label the net force vector at point 2.

4.7 Free-Body Diagrams

Exercises 17–22:
- Draw a picture and identify the forces, then
- Draw a complete free-body diagram for the object, following each of the steps given in Tactics Box 4.3. Be sure to think carefully about the direction of $\vec{F}_{net}$.

Note: Draw individual force vectors with a **black** or **blue** pencil or pen. Draw the *net* force vector $\vec{F}_{net}$ with a **red** pencil or pen.

17. A heavy crate is being lowered straight down at a constant speed by a steel cable.

18. A boy is pushing a box across the floor at a steadily increasing speed. Let the box be "the system" for analysis.

19. A bicycle is speeding up down a hill. Friction is negligible, but air resistance is not.

20. You've slammed on your car brakes while going down a hill. You're skidding to a halt.

21. You are going to toss a rock *straight up* into the air by placing it on the palm of your hand (you're not gripping it), then pushing your hand up very rapidly. You may want to toss an object into the air this way to help you think about the situation. The rock is "the system" of interest.

    a. As you hold the rock at rest on your palm, before moving your hand.

    b. As your hand is moving up but before the rock leaves your hand.

    c. One-tenth of a second after the rock leaves your hand.

    d. After the rock has reached its highest point and is now falling straight down.

22. Block B has just been released and is beginning to fall. Consider block A to be "the system."

4.8 Newton's Third Law

Exercises 23–27: Apply the steps of Tactics Box 4.4 for identifying forces for interacting systems for each situation. That is,
- Identify the relevant objects in each case and draw a picture showing each relevant object separate from all other objects, but in the correct spatial orientation. Include the earth and, if appropriate, the earth's surface.
- Identify and label all forces as appropriate on the objects using the techniques of Tactics Box 4.2.
- Identify and label all action/reaction pairs using the notation $\vec{F}_{\text{A on B}}$.
- Draw a separate free-body diagram for each object and show the forces as **black** vectors.
- Connect all action/reaction pairs on the free-body diagrams with **red** dotted lines.

Note: Your pictures should look similar to Figure 4.34 in your text.

23. A boy pulls a wagon by a rope attached to the front of the wagon. The rope is parallel to the ground. Rolling friction is not negligible.

24. A bicycle accelerates forward from rest. (Treat the bicycle and its rider as a single object.)

25. a. A bat hits a ball. (Draw your picture from the perspective of someone seeing the *end* of the bat at the moment it strikes the ball.)

    b. The ball then sails through the air.

26. a. A ball hangs from a string. The string is attached to the ceiling.

    b. The string is cut. The ball falls and bounces. Consider the instant that the ball is in contact with the ground. (You don't need to show the string or the ceiling in part b.)

27. A crate is in the back of a truck as the truck accelerates forward. The crate does not slip. (Treat the crate and the truck as separate systems.)

28. You find yourself in the middle of a frozen lake with a surface so slippery ($\mu_s = \mu_k = 0$) that you cannot walk. However, you happen to have several rocks in your pocket. The ice is extremely hard. It cannot be chipped, and the rocks slip on it just as much as your feet do. Can you think of a way to get to shore? Use pictures, forces, and Newton's laws to explain your reasoning.

29. How do you paddle a canoe in the forward direction? Explain. Your explanation should include pictures showing forces on the water and forces on the paddle.

30. When you blow up a balloon and release it, it shoots forward. Explain why. Include pictures showing forces on the balloon and forces on the parcel of air that was just expelled from the balloon.

31. How does a rocket take off? What is the upward force on it? Your explanation should include pictures showing forces on the rocket and forces on the parcel of hot gas that was just expelled from the rocket's exhaust.

32. How do basketball players jump straight up into the air? Your explanation should include pictures showing forces on the player and forces on the ground.

5 Applying Newton's Laws

5.1 Equilibrium

1. If an object is at rest, can you conclude that there are no forces acting on it? Explain.

2. If a force is exerted on an object, is it possible for that object to be moving with constant velocity? Explain.

3. A hollow tube forms three-quarters of a circle. It is lying flat on a table. A ball is shot through the tube at high speed. As the ball emerges from the other end, does it follow path A, path B, or path C? Explain your reasoning.

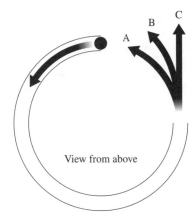

View from above

4. The vectors below show five forces that can be applied individually or in combinations to an object. Which forces or combinations of forces will cause the object to be in equilibrium?

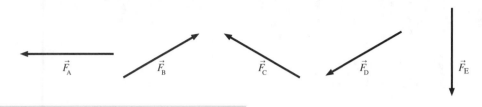

5. The free-body diagrams show a force or forces acting on an object. Draw and label one more force (one that is appropriate to the situation) that will cause the object to be in equilibrium.

6. The free-body diagrams show a force or forces acting on an object. Draw and label one more force (one that is appropriate to the situation) that will cause the object to be in equilibrium.

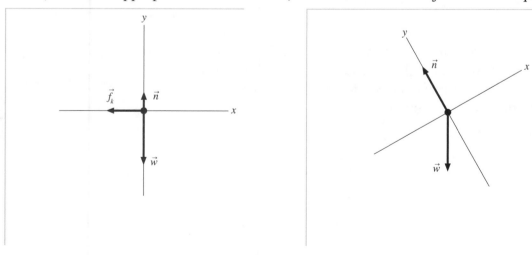

5.2 Dynamics and Newton's Second Law

7. a. An elevator travels *upward* at a constant speed. The elevator hangs by a single cable. Friction and air resistance are negligible. Is the tension in the cable greater than, less than, or equal to the weight of the elevator? Explain. Your explanation should include both a free-body diagram and reference to appropriate physical principles.

   b. The elevator travels *downward* and is slowing down. Is the tension in the cable greater than, less than, or equal to the weight of the elevator? Explain.

Exercises 8–9: The figures show free-body diagrams for an object of mass m. Write the x- and y-components of Newton's second law. Write your equations in terms of the *magnitudes* of the forces $F_1, F_2, \ldots$ and any *angles* defined in the diagram. One equation is shown to illustrate the procedure.

8.

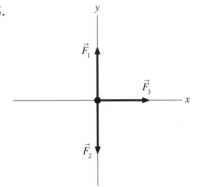

$ma_x =$

$ma_y = F_1 - F_2$

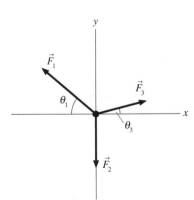

$ma_x =$

$ma_y =$

9.

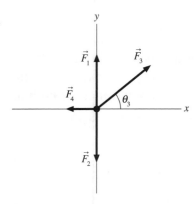
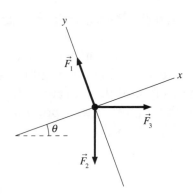

$ma_x = F_3 \cos\theta_3 - F_4$	$ma_x =$
$ma_y =$	$ma_y =$

Exercises 10–12: Two or more forces, shown on a free-body diagram, are exerted on a 2 kg object. The units of the grid are newtons. For each:

- Draw a vector arrow *on the grid,* starting at the origin, to show the net force $\vec{F}_{net}$.
- In the space to the right, determine the numerical values of the components a_x and a_y.

10.

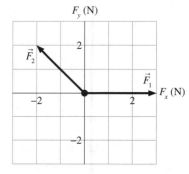

$a_x =$

$a_y =$

11.

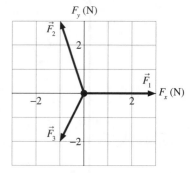

$a_x =$

$a_y =$

12.

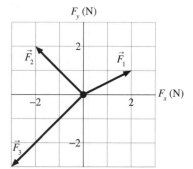

$a_x =$

$a_y =$

Exercises 13–15: Three forces $\vec{F}_1$, $\vec{F}_2$, and $\vec{F}_3$ cause a 1 kg object to accelerate with the acceleration given. Two of the forces are shown on the free-body diagrams below, but the third is missing. For each, draw and label *on the grid* the missing third force vector.

13. $\vec{a} = (2\,\text{m/s}^2)$

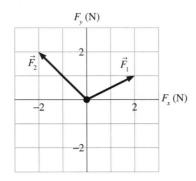

14. $\vec{a} = (0, -3\,\text{m/s}^2)$

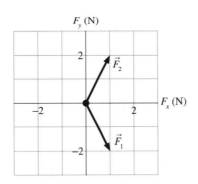

15. The object moves with constant velocity.

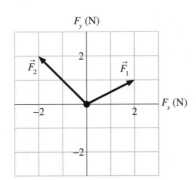

5.3 Mass and Weight

16. Suppose you have a jet-powered flying platform that can move straight up and down. For each of the following cases, is your apparent weight equal to, greater than, or less than your true weight? Explain.

    a. You are ascending and speeding up.

    b. You are descending and speeding up.

    c. You are ascending at a constant speed.

    d. You are ascending and slowing down.

    e. You are descending and slowing down.

17. The terms "vertical" and "horizontal" are frequently used in physics. Give *operational definitions* for these two terms. An operational definition defines a term by how it is measured or determined. Your definition should apply equally well in a laboratory or on a steep mountainside.

18. An astronaut orbiting the earth is handed two balls that are identical in outward appearance. However, one is hollow while the other is filled with lead. How might the astronaut determine which is which? Cutting them open is not allowed.

5.4 Normal Forces

19. Suppose you stand on a spring scale in six identical elevators. Each elevator moves as shown below. Let the reading of the scale in elevator n be S_n. Rank in order, from largest to smallest, the six scale readings S_1 to S_6. Some may be equal. Give your answer in the form $A > B = C > D$.

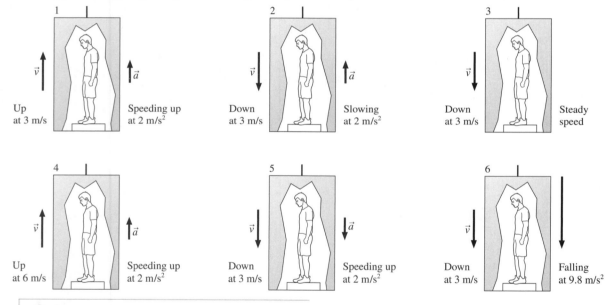

Order:

Explanation:

5.5 Friction

20. Suppose you press a book against the wall with your hand. The book is not moving.

   a. Identify the forces on the book and draw a free-body diagram.

   b. Now suppose you decrease your push, but not enough for the book to slip. What happens to each of the following forces? Do they increase in magnitude, decrease, or not change?

   $\vec{F}_{\text{push}}$ _____

   $\vec{w}$ _____

   $\vec{n}$ _____

   $\vec{f}_{\text{s}}$ _____

   $f_{\text{s max}}$ _____

21. Consider a box in the back of a pickup truck.

   a. If the truck accelerates slowly, the box moves with the truck without slipping. What force or forces act on the box to accelerate it? In what direction do those forces point?

   b. Draw a free-body diagram of the box.

   c. What happens to the box if the truck accelerates too rapidly? *Explain* why this happens, basing your explanation on physical models and the principles described in this chapter.

5.6 Drag

22. Five balls move through the air as shown. All five have the same size and shape. Rank in order, from largest to smallest, the size of their accelerations a_1 to a_5. Some may be equal. Give your answer in the form A > B = C > D.

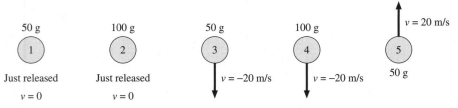

Order:

Explanation:

23. A 1 kg wood ball and a 10 kg lead ball have identical shapes and sizes. They are dropped simultaneously from a tall tower.

a. To begin, assume that air resistance is negligible. As the balls fall, are the forces on them equal in magnitude or different? If different, which has the larger force? _____

b. Are their accelerations equal? If different, which has the larger acceleration? Explain.

c. Which ball hits the ground first? Or do they hit simultaneously? Explain.

d. When air resistance is included, each ball will experience the *same* drag force when moving at the same speed because both have the same shape. Are the accelerations of the balls now equal or different? If not, which has the larger acceleration? Explain, using your free-body diagrams and Newton's laws.

e. Which ball now hits the ground first? Or do they hit simultaneously? Explain.

5.7 Interacting Objects

24. Block A is pushed across a horizontal surface at a *constant* speed by a hand that exerts force $\vec{F}_{\text{H on A}}$. The surface has friction.

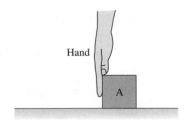

Hand

A

   a. Draw two free-body diagrams, one for the hand and the other for the block. On these diagrams, show only the *horizontal* forces with lengths portraying the relative magnitudes of the forces. Label force vectors, using the form $\vec{F}_{\text{C on D}}$. Connect action/reaction pairs with dotted lines. On the hand diagram show only $\vec{F}_{\text{H on A}}$. Don't include $\vec{F}_{\text{body on H}}$.

   b. Rank in order, from largest to smallest, the magnitudes of *all* of the horizontal forces you showed in part a. For example, if $F_{\text{C on D}}$ is the largest of three forces while $F_{\text{D on C}}$ and $F_{\text{D on E}}$ are smaller but equal, you can record this as $F_{\text{C on D}} > F_{\text{D on C}} = F_{\text{D on E}}$.

   Order:

   Explanation:

25. A second block B is placed in front of Block A of question 24. B is more massive than A: $m_{\text{B}} > m_{\text{A}}$. The blocks are speeding up.

   a. Consider a *frictionless* surface. Draw *separate* free-body diagrams for A, B, and the hand. Show only the horizontal forces. Label forces in the form $\vec{F}_{\text{C on D}}$. Use dotted lines to connect action/reaction pairs.

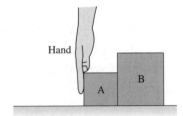

Hand

A   B

   b. By applying the second law to each block and the third law to each action/reaction pair, rank in order *all* of the horizontal forces, from largest to smallest.

   Order:

   Explanation:

26. Blocks A and B are held on the palm of your outstretched hand as you lift them straight up at *constant speed*. Assume $m_B > m_A$ and that $m_{hand} = 0$.

    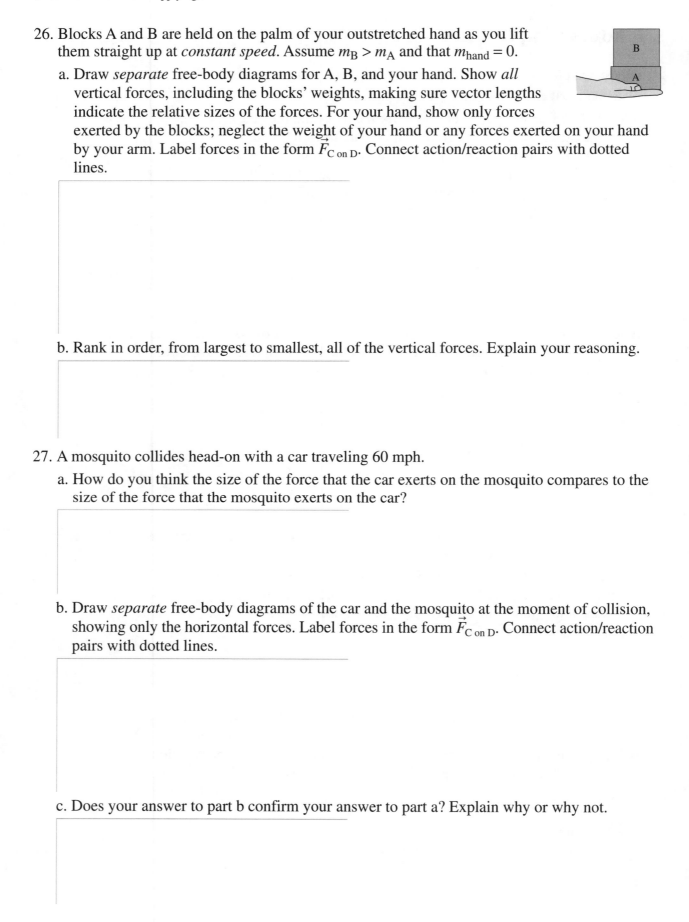

    a. Draw *separate* free-body diagrams for A, B, and your hand. Show *all* vertical forces, including the blocks' weights, making sure vector lengths indicate the relative sizes of the forces. For your hand, show only forces exerted by the blocks; neglect the weight of your hand or any forces exerted on your hand by your arm. Label forces in the form $\vec{F}_{C\ on\ D}$. Connect action/reaction pairs with dotted lines.

    b. Rank in order, from largest to smallest, all of the vertical forces. Explain your reasoning.

27. A mosquito collides head-on with a car traveling 60 mph.

    a. How do you think the size of the force that the car exerts on the mosquito compares to the size of the force that the mosquito exerts on the car?

    b. Draw *separate* free-body diagrams of the car and the mosquito at the moment of collision, showing only the horizontal forces. Label forces in the form $\vec{F}_{C\ on\ D}$. Connect action/reaction pairs with dotted lines.

    c. Does your answer to part b confirm your answer to part a? Explain why or why not.

5.8 Ropes and Pulleys

28. Blocks A and B are connected by a massless string over a massless, frictionless pulley. The blocks have just this instant been released from rest.

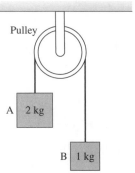

a. Will the blocks accelerate? If so, in which directions?

b. Draw a separate free-body diagram for each block. Be sure vector lengths indicate the relative size of the force. Connect any action/reaction pairs or "forces that act *as if* they are action/reaction" pairs with dotted lines.

c. Rank in order, from largest to smallest, all of the vertical forces. Explain.

d. Compare the magnitude of the *net* force on A with the *net* force on B. Are they equal, or is one larger than the other? Explain.

e. Consider the block that falls. Is the magnitude of its acceleration less than, greater than, or equal to *g*? Explain.

29. In case a, block A is accelerated across a frictionless table by a hanging 10 N weight (1.02 kg). In case b, the same block is accelerated by a steady 10 N tension in the string.

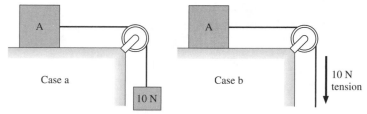

Case a

Case b

10 N tension

Is block A's acceleration in case b greater than, less than, or equal to its acceleration in case a? Explain.

Exercises 30–31: Draw separate free-body diagrams for blocks A and B. Connect any action/reaction pairs (or forces that act *as if* they are action/reaction pairs) together with dotted lines.

30.

31.

6 Circular Motion, Orbits, and Gravity

6.1 Uniform Circular Motion

6.2 Speed, Velocity and Acceleration in Uniform Circular Motion

1. On each of the two circles shown below, draw a motion diagram for an object moving at a constant angular speed of $\omega = +0.5$ rad/s along the perimeter starting at point P and through an angle of *two* radians. Indicate the location of the particle every 1.0 s. Draw velocity vectors **black** and acceleration vectors **red.**

a.
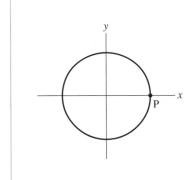

b.
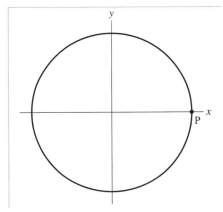

Plot the angle as a function of time for each of two motion diagrams that you have drawn above. Include an appropriate scale for the vertical axis in each case.

c.
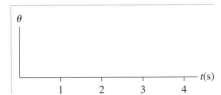

d.
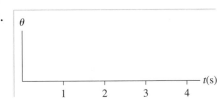

2. The figure shows three points on a steadily rotating wheel.

Rank in order, from largest to smallest, the angular velocities ω_1, ω_2, and ω_3 of these points.

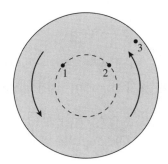

Order:

Explanation:

3. Below are two angular position-versus-time graphs. For each, draw the corresponding angular velocity-versus-time graph directly below it.

a.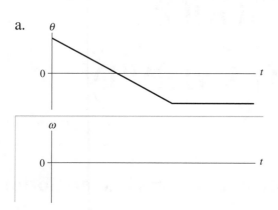

b.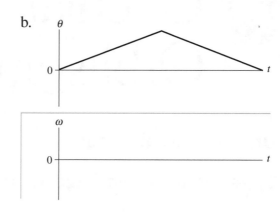

4. Below are two angular velocity-versus-time graphs. For each, draw the corresponding angular position-versus-time graph directly below it. Assume $\theta_0 = 0$ rad.

a.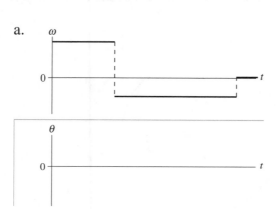

b.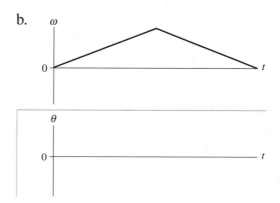

5. A particle in circular motion rotates clockwise at 4 rad/s for 2 s, then counterclockwise at 2 rad/s for 4 s. The time required to change direction is negligible. Graph the angular velocity and the angular position, assuming $\theta_0 = 0$ rad.

 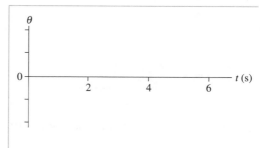

6. A particle travels at constant speed along a circle with $a = 8$ m/s². What is a if

a. The radius is doubled without changing the angular velocity? _____

b. The radius is doubled without changing the particle's speed? _____

c. The angular velocity is doubled without changing the particle's radius? _____

6.3 Dynamics of Uniform Circular Motion

7. The figure shows a *top view* of a plastic tube that is fixed on a horizontal table top. A marble is shot into the tube at A. Sketch the marble's trajectory after it leaves the tube at B. Explain.

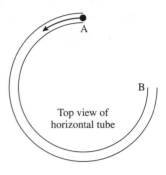

Top view of horizontal tube

8. A ball swings in a *vertical* circle on a string. During one revolution, a very sharp knife is used to cut the string at the instant when the ball is at its lowest point. Sketch the subsequent trajectory of the ball until it hits the ground. Explain.

Knife

9. The figures are a bird's-eye view of particles moving in horizontal circles on a table top. All are moving at the same speed. Rank in order, from largest to smallest, the tensions T_1 to T_4.

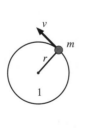

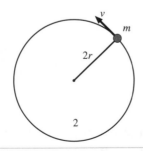

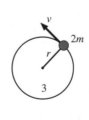

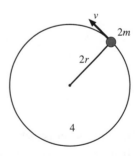

Order:

Explanation:

10. A ball on a string moves in a vertical circle. When the ball is at its lowest point, is the tension in the string greater than, less than, or equal to the ball's weight? Explain. (You should include a free-body diagram as part of your explanation.)

11. A marble rolls around the inside of a cone. Draw a free-body diagram of the marble when it is on the left side of the cone and a free-body diagram of the marble when it is on the right side of the cone.

On left side   On right side

12. A jet airplane is flying on a level course at constant velocity.

    a. What is the *net* force on the plane? _____

    b. Draw a free-body diagram and identify all of the forces acting on the plane.

    c. Airplanes bank when they turn. Explain why, in terms of forces and physical laws.
       Hint: What would a free-body diagram look like to an observer *behind* the plane?

6.4 Apparent Forces in Circular Motion

13. The drawing shows a car moving clockwise at constant speed over the top of a circular hill.
    a. On the drawing, complete the motion diagram by showing the car's velocity vectors for each of the three motion diagram points and use these to indicate the acceleration of the car at the top of the hill.
    b. To the right of the sketch, draw a free-body diagram for the car when at the top of the hill and indicate the direction of the net force on the car.

    c. Is your free-body diagram consistent with the motion diagram? Explain.

    d. For this situation, is there a maximum speed at which the car can travel over the top of the hill and not lose contact with the hill? If not, why not? If so, show how your free-body diagram would change, if at all, at that speed.

14. The drawing shows a car moving upside down while looping a circular roller coaster loop-the-loop at constant speed in the clockwise direction.

    a. On the drawing, complete the motion diagram by showing the car's velocity vectors for each of the three motion diagram points and use these to indicate the acceleration of the car at the top of the loop-the-loop.

    b. To the right of the sketch, draw a free-body diagram for the car at the top of the loop and indicate the direction of the net force on the car. (Assume the car is moving fast enough so that it would not fall, even if not attached to the roller coaster track.)

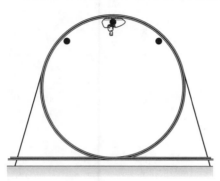

    c. Is your free-body diagram consistent with your motion diagram? Explain.

    d. For this situation, is there a minimum speed at which the car can travel over the top of the loop and not lose contact with the loop? If not, why not? If so, show how your free-body diagram would change, if at all, at that speed.

6.5 Circular Orbits and Weightlessness

15. The earth has seasons because the axis of the earth's rotation is tilted 23° away from a line perpendicular to the plane of the earth's orbit. You can see this in the figure, which shows the edge of the earth's orbit around the sun. For both positions of the earth, draw a force vector to show the net force acting on the earth or, if appropriate, write $\vec{F} = \vec{0}$.

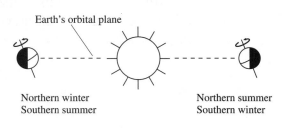

Earth's orbital plane

Northern winter
Southern summer

Northern summer
Southern winter

16. A small projectile is launched parallel to the ground at height $h = 1$ m with sufficient speed to orbit a completely smooth, airless planet. A bug rides in a small hole inside the projectile. Is the bug weightless? Explain.

17. A stunt plane does a series of vertical loop-the-loops. At what point in the circle does the pilot feel the heaviest? Explain. Include a free-body diagram with your explanation.

18. A roller-coaster car goes around the inside of a loop-the-loop. Check the statement that is true when the car is at the highest point and at the lowest point in the loop.

	Highest	Lowest
The apparent weight w_{app} is always less than w		
The apparent weight w_{app} is always equal to w		
The apparent weight w_{app} is always greater than w		
w_{app} could be less than, equal to, or greater than w		

19. You can swing a ball on a string in a *vertical* circle if you swing it fast enough.

    a. Draw two free-body diagrams of the ball at the top of the circle. On the left, show the ball when it is going around the circle very fast. On the right, show the ball as it goes around the circle more slowly.

Very fast	Slower

    b. As you continue slowing the swing, there comes a frequency at which the string goes slack and the ball doesn't make it to the top of the circle. What condition must be satisfied for the ball to be able to complete the full circle?

    c. Suppose the ball has the smallest possible frequency that allows it to go all the way around the circle. What is the tension in the string when the ball is at the highest point? Explain.

20. It's been proposed that future space stations create "artificial gravity" by rotating around an axis. (The space station would have to be much larger than the present space station for this to be feasible.)

    a. How would this work? Explain.

    b. Would the artificial gravity be equally effective throughout the space station? If not, where in the space station would the residents want to live and work?

6.6 Newton's Law of Gravity

21. Is the earth's gravitational force on the sun larger than, smaller than, or equal to the sun's gravitational force on the earth? Explain.

22. Star A is twice as massive as star B.

    a. Draw gravitational force vectors on both stars. The length of each vector should be proportional to the size of the force.

    $m_A = 2m_B$ m_B

    b. Is the acceleration of star A larger than, smaller than, or equal to the acceleration of star B? Explain.

23. The quantity y is inversely proportional to the square of x and $y = 4$ when $x = 5$.

    a. Write an equation to represent this inverse-square relationship for all y and x.

    b. Find y if $x = 2$. _____   c. Find x if $y = 100$. _____

    d. By what factor must x change for the value of y to double? _____

    e. Compare your equation in part a to the equation from your text relating the force of gravitational attraction of two objects F to the distance between them r,

    $$F_{1 \text{ on } 2} = F_{2 \text{ on } 1} = \frac{Gm_1m_2}{r^2}.$$

    Which quantity assumes the role of x? _____

    Which quantity assumes the role of y? _____

    What is the constant of proportionality relating F and r^2? _____

24. How far away from the earth does an orbiting spacecraft have to be in order for the astronauts inside to be "weightless"?

25. The acceleration due to gravity at the surface of planet 1 is 20 m/s². The radius and the mass of planet 2 are twice those of planet 1. What is g on planet 2?

6.7 Gravity and Orbits

26. Planet X orbits the star Omega with a "year" that is 200 earth days long. Planet Y circles Omega at four times the distance of planet X. How long is a year on planet Y?

27. The mass of Jupiter is $M_{Jupiter} = 300M_{earth}$. Jupiter orbits around the sun with $T_{Jupiter} = 11.9$ years in an orbit with $r_{Jupiter} = 5.2r_{earth}$. Suppose the earth could be moved to the distance of Jupiter and placed in a circular orbit around the sun. The new period of the earth's orbit would be

    a. 1 year.   b. 11.9 years.

    c. Between 1 year and 11.9 years.   d. More than 11.9 years.

    e. It could be anything, depending on   f. It is impossible for a planet of earth's mass
       the speed the earth is given.   to orbit at the distance of Jupiter.

    Circle the letter of the true statement. Then explain your choice.

28. The gravitational force of a star on orbiting planet 1 is F_1. Planet 2, which is twice as massive as planet 1 and orbits at twice the distance from the star, experiences gravitational force F_2.

    a. What is the ratio F_2/F_1?

    b. Planet 1 orbits the star with period T_1 and planet 2 with period T_1. What is the ratio T_2/T_1?

29. Satellite A orbits a planet with a speed of 10,000 m/s. Satellite B is twice as massive as satellite A and orbits at twice the distance from the center of the planet. What is the speed of satellite B?

7 Rotational Motion

7.1 The Rotation of a Rigid Body

1. A pendulum swings from its end point on the left (point 1) to its end point on the right (point 5). At each of the labeled points:

   a. Use a **black** pen or pencil to draw and label the vectors $\vec{a}_c$ and $\vec{a}_t$ at each point. Make sure the length indicates the relative size of the vector.

   b. Use a **red** pen or pencil to draw and label the total acceleration vector $\vec{a}$.

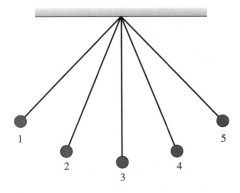

2. The following figures show a rotating wheel. If we consider a counterclockwise rotation as positive (+) and a clockwise rotation as negative (−), determine the signs (+ or −) of ω and α.

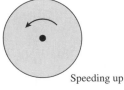 Speeding up

 Slowing down

 Slowing down

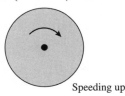 Speeding up

ω _____ ω _____ ω _____ ω _____

α _____ α _____ α _____ α _____

3. The figures below show the centripetal acceleration vector $\vec{a}_c$ at four successive points on the trajectory of a particle moving in a counterclockwise circle.

   a. For each, draw the tangential acceleration vector $\vec{a}_t$ at points 2 and 3 or, if appropriate, write $\vec{a}_t = \vec{0}$.

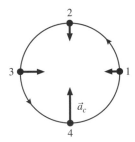   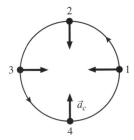   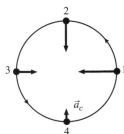

   b. If we consider a counterclockwise rotation as positive and a clockwise rotation as negative, determine if the particle's angular acceleration α is positive (+), negative (−), or zero (0).

   $\alpha =$ _____ $\alpha =$ _____ $\alpha =$ _____

4. Below are three angular velocity-versus-time graphs. For each, draw the corresponding angular acceleration-versus-time graph.

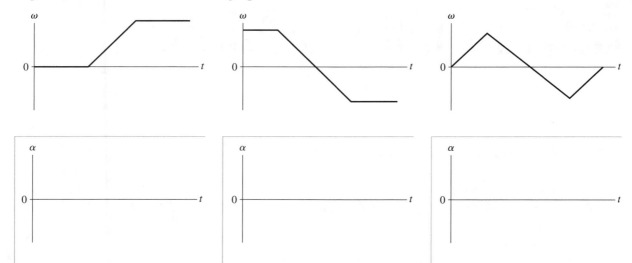

5. A wheel rolls to the left along a horizontal surface, down a ramp, then continues along the lower horizontal surface. Draw graphs for the wheel's angular velocity ω and angular acceleration α as functions of time.

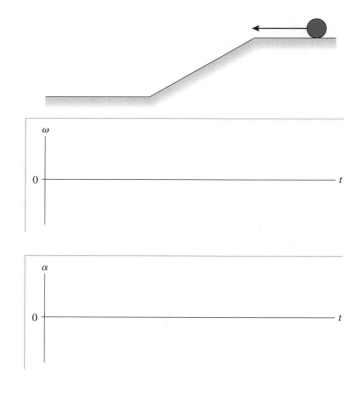

7.2 Torque

6. Five forces are applied to a door. For each, determine if the torque about the hinge is positive (+), negative (−), or zero (0).

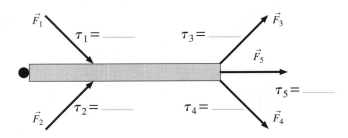

$\tau_1 =$ _____ $\tau_3 =$ _____

$\tau_5 =$ _____

$\tau_2 =$ _____ $\tau_4 =$ _____

7. Six forces, each of magnitude either F or $2F$, are applied to a door. Rank in order, from largest to smallest, the six torques τ_1 to τ_6 about the hinge.

Order:

Explanation:

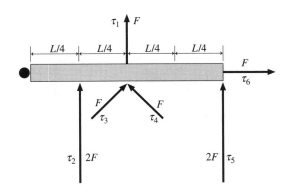

8. Four forces are applied to a rod that can pivot on an axle. For each force,
   a. Use a **black** pen or pencil to draw the line of action.
   b. Use a **red** pen or pencil to draw and label the moment arm, or state that $d = 0$.
   c. Determine if the torque about the axle is positive (+), negative (−), or zero (0).

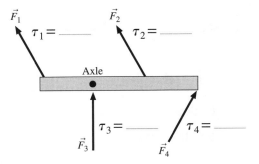

$\tau_1 =$ _____ $\tau_2 =$ _____

$\tau_3 =$ _____ $\tau_4 =$ _____

9. a. Draw a force vector at A whose torque about the axle is negative.
   b. Draw a force vector at B whose torque about the axle is zero.
   c. Draw a force vector at C whose torque about the axle is positive.

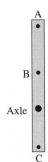

10. The dumbbells below are all the same size, and the forces all have the same magnitude. Rank in order, from largest to smallest, the torques τ_1, τ_2, and τ_3 about the midpoint of each connecting rod.

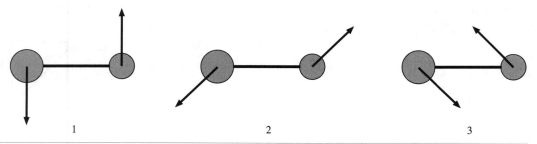

Order:

Explanation:

11. a. Rank in order, from largest to smallest, the torques τ_1 to τ_4.

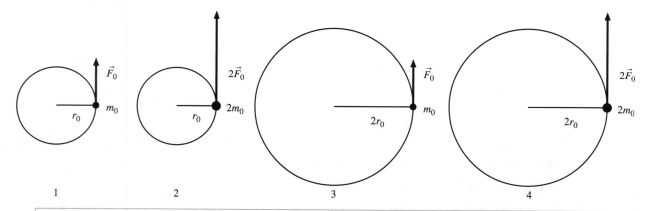

Order:

Explanation:

b. Rank in order, from largest to smallest, the angular accelerations α_1 to α_4.

Order:

Explanation:

7.3 Gravitational Torque and the Center of Gravity

12. a. Find the coordinates for and *mark* the center of gravity for the pair of masses shown, using the center of the 3.0 kg mass as the origin.

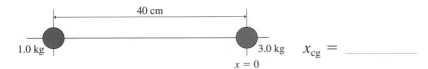

$x_{cg} =$ _____

b. Find the coordinates for and *mark* the center of gravity for the pair of masses shown, using the center of the 1.0 kg mass as the origin.

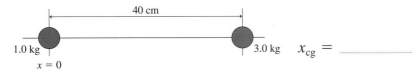

$x_{cg} =$ _____

c. How do the locations for the marks in parts a and b compare? How do the coordinates compare?

d. The 3.0 kg mass from parts a and b above is separated into two 1.5 kg pieces. One of these is moved 10 cm in the +y-direction. Find the coordinates for and *mark* the center of gravity of the new system using the origin at the left-side mass.

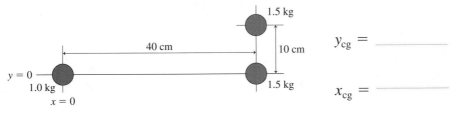

$y_{cg} =$ _____

$x_{cg} =$ _____

e. What effect did separating the 3.0 kg mass along the y-direction have on the x-component of the center of gravity of the system?

13. Is the center of gravity of this dumbbell at point 1, 2, or 3? Explain. (Assume the end masses are of the same material.)

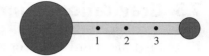

14. Mark the center of gravity of this object with an ×. Explain.

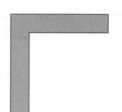

7.4 Rotational Dynamics and Moment of Inertia

15. A student gives a quick push to a ball at the end of a massless, rigid rod, causing the ball to rotate clockwise in a *horizontal* circle. The rod's pivot is frictionless.

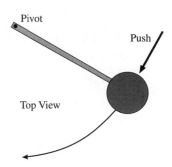

a. As the student is pushing, is the torque about the pivot positive, negative, or zero?_____

b. After the push has ended, does the ball's angular velocity

   i.   Steadily increase?

   ii. Increase for awhile, then hold steady?

   iii. Hold steady?

   iv. Decrease for awhile, then hold steady?

   v. Steadily decrease?

Explain the reason for your choice.

c. Right after the push has ended, is the torque positive, negative, or zero?_____

16. The top graph shows the torque on a rotating wheel as a function of time. The wheel's moment of inertia is 10 kg m^2. Draw graphs of α-versus-t and ω-versus-t, assuming $\omega_0 = 0$.

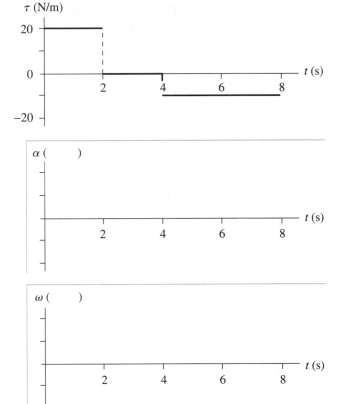

17. The wheel turns on a frictionless axle. A string wrapped around the smaller diameter shaft is tied to a block. The block is released at $t = 0$ s and hits the ground at $t = t_1$.

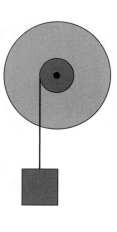

a. Draw a graph of ω-versus-t for the wheel, starting at $t = 0$ s and continuing to some time $t > t_1$.

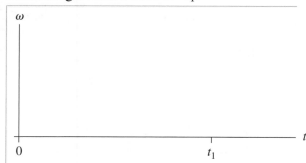

b. Is the magnitude of the block's downward acceleration greater than g, less than g, or equal to g? Explain.

18. The moment of inertia of a uniform rod about an axis through its center is $\frac{1}{12} ML^2$. The moment of inertia about an axis at one end is $\frac{1}{3} ML^2$. Explain *why* the moment of inertia is larger about the end than about the center.

19. You have two steel spheres. Sphere 2 has twice the radius of sphere 1. By what *factor* does the moment of inertia I_2 of sphere 2 exceed the moment of inertia I_1 of sphere 1?

20. The professor hands you two spheres. They have the same mass, the same radius, and the same exterior surface. The professor claims that one is a solid sphere and that the other is hollow. Can you determine which is which without cutting them open? If so, how? If not, why not?

21. Rank in order, from largest to smallest, the moments of inertia I_1, I_2, and I_3 about the midpoint of each connecting rod.

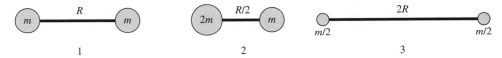

Order:

Explanation:

7.5 Using Newton's Second Law for Rotation

22. A square plate can rotate about an axle through its center. Four forces of equal magnitude are applied to different points on the plate. The forces turn as the plate rotates, maintaining the same orientation with respect to the plate. Rank in order, from largest to smallest, the angular accelerations α_1 to α_4.

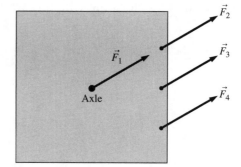

Order:

Explanation:

23. A solid cylinder and a cylindrical shell have the same mass, same radius, and turn on frictionless, horizontal axles. (The cylindrical shell has lightweight spokes connecting the shell to the axle.) A rope is wrapped around each cylinder and tied to a block. The blocks have the same mass and are held the same height above the ground. Both blocks are released simultaneously. The ropes do not slip.

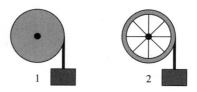

Which block hits the ground first? Or is it a tie? Explain.

7.6 Rolling Motion

24. A wheel is rolling along a horizontal surface with the center-of-mass velocity shown. Draw the velocity vector $\vec{v}$ at points 1 to 4 on the rim of the wheel.

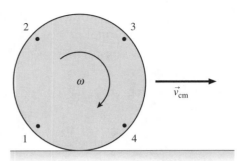

25. A wheel is rolling along a horizontal surface with the center-of-mass velocity shown. Draw the velocity vector $\vec{v}$ at points 1 to 3 on the wheel.

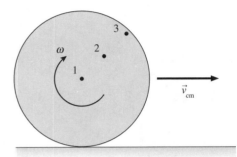

26. If a solid disk and a circular hoop of the same mass and radius are released from rest at the top of a ramp and allowed to roll to the bottom, the disk will get to the bottom first. *Without referring to equations*, explain why this is so.

8 Equilibrium and Elasticity

8.1 Torque and Static Equilibrium

1. The dumbbell has masses m and $2m$. Force $\vec{F_1}$ acts on mass m in the direction shown. Is there a force $\vec{F_2}$ that can act on mass $2m$ such that the dumbbell moves with pure translational motion, without any rotation? If so, draw $\vec{F_2}$, making sure that its length shows the magnitude of $\vec{F_2}$ relative to $\vec{F_1}$. If not, explain why not.

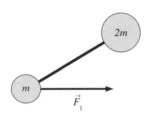

2. Forces $\vec{F_1}$ and $\vec{F_2}$ have the same magnitude and are applied to the corners of a square plate. Is there a *single* force $\vec{F_3}$ that, if applied to the appropriate point on the plate, will cause the plate to be in total equilibrium? If so, draw it, making sure it has the right position, orientation, and length. If not, explain why not.

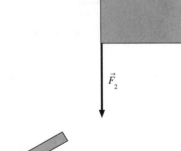

3. A uniform rod pivots about a frictionless, horizontal axle through its center. It is placed on a stand, held motionless in the position shown, then gently released. On the right side of the figure, draw the final, equilibrium position of the rod. Explain your reasoning.

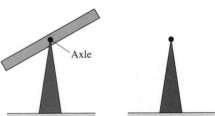

Exercises 4–8: For each of the following
- Draw and label the appropriate force vectors at the locations at which each force acts on the beams and with the appropriate relative lengths so that each beam is in equilibrium. Assume each beam is uniform and has a weight w. Assume each block also has a weight w.
- Write three equilibrium equations to sum the vertical forces, the horizontal forces, and the torques, respectively, on each beam so that each beam is can be shown to be in equilibrium.

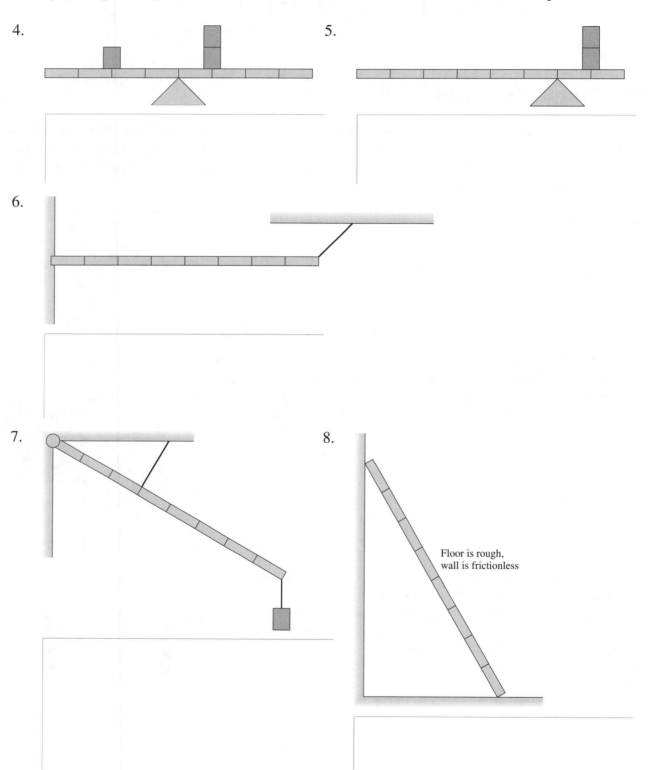

4.

5.

6.

7.

8.

Floor is rough, wall is frictionless

8.2 Stability and Balance

9. The center of gravity (⊙) is marked on each of the objects shown below.
   - Mark and label the track width supporting the object.
   - Clearly label the critical angle for each object.
   - Measure the height to the center of gravity and the track width and determine the approximate value for the critical angle in each case.

a.

b.

c.

$\theta_c =$

$\theta_c =$

$\theta_c =$

10. Tight rope walkers often carry long poles that are weighted at the ends. These poles serve at least two purposes, one is to change the critical angle and the second is to increase the walker's moment of inertia.

    a. Use the diagram below to describe how the critical angle is changed by the tightrope walker's use of the pole.

    b. Why would increasing the tightrope walker's moment of inertia also help to make him less likely to fall?

8.3 Springs and Elastic Materials

11. A spring is attached to the floor and pulled straight up by a string. The spring's tension is measured. The graph shows the tension in the spring as a function of the spring's length L.

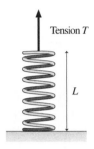
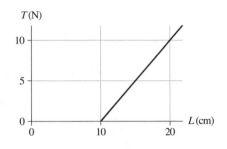

a. Does this spring obey Hooke's Law? Explain why or why not.

b. If it does, what is the spring constant?

12. A spring has an unstretched length of 10 cm. It exerts a restoring force F when stretched to a length of 11 cm.

a. For what length of the spring is its restoring force $3F$? _____

b. At what compressed length is the restoring force $2F$? _____

13. The left end of a spring is attached to a wall. When Bob pulls on the right end with a 200 N force, he stretches the spring by 20 cm. The same spring is then used for a tug-of-war between Bob and Carlos. Each pulls on his end of the spring with a 200 N force.

a. How far does Bob's end of the spring move? Explain.

b. How far does Carlos's end of the spring move? Explain.

14. The graph below shows the stretching of two different springs, A and B, when different forces were applied.

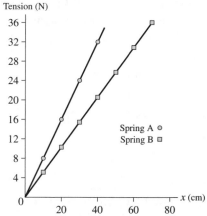

a. Determine the spring constant for each spring.

$k_A =$ _____

$k_B =$ _____

b. For each spring, determine the amount of tension required to stretch the spring 40.0 cm and mark that tension on the graph.

$T_A =$ _____

$T_B =$ _____

c. Can you determine the amount of tension required to stretch each spring by 400 cm? If so, what is the tension required in each case. If not, why not?

8.4 Stretching, Compressing and Bending Materials

15. A force stretches a wire by 1 mm.
    a. A second wire of the same material has the same cross section and twice the length. How far will it be stretched by the same force? Explain.

    b. A third wire of the same material has the same length and twice the diameter as the first. How far will it be stretched by the same force? Explain.

16. A 2000 N force stretches a wire by 1 mm.
    a. A second wire of the same material is twice as long and has twice the diameter. How much force is needed to stretch it by 1 mm? Explain.

    b. A third wire is twice as long as the first and has the same diameter. How far is it stretched by a 4000 N force?

17. A wire is stretched right to the breaking point by a 5000 N force. A longer wire made of the same material has the same diameter. Is the force that will stretch it right to the breaking point larger than, smaller than, or equal to 5000 N? Explain.

9 Momentum

9.1 Impulse

1. The position-versus-time graph is shown for a 500 g object. Draw the corresponding momentum-versus-time graph. Include an appropriate vertical scale.

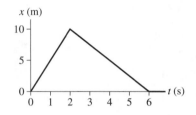

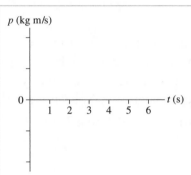

2. The momentum-versus-time graph is shown for a 500 g object. Draw the corresponding acceleration-versus-time graph. Include an appropriate vertical scale.

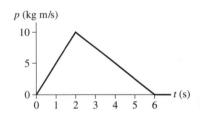

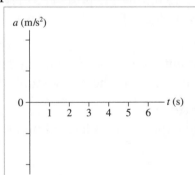

3. A 2 kg object is moving to the right with a speed of 1 m/s when it experiences an impulse due to the force shown in the graph. What is the object's speed and direction after the impulse?

a.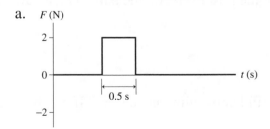

b.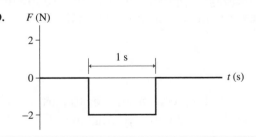

4. A 2 kg object is moving to the right with a speed of 1 m/s when it experiences an impulse due to the force shown in the graph. What is the object's speed and direction after the impulse?

a.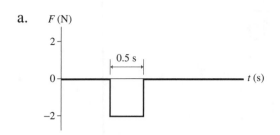

b.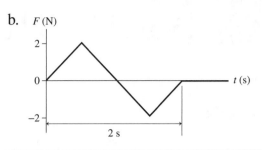

5. A carnival game requires you to knock over a wood post by throwing a ball at it. You're offered a very bouncy rubber ball and a very sticky clay ball of equal mass. Assume that you can throw them with equal speed and equal accuracy. You only get one throw.

   a. Which ball will you choose? Why?

   b. Let's think about the situation more carefully. Both balls have the same initial momentum p_{ix} just before hitting the post. The clay ball sticks, the rubber ball bounces off with essentially no loss of speed. What is the final momentum of each ball?

   Clay ball: p_{fx} = _____   Rubber ball: p_{fx} = _____
   Hint: Momentum has a sign. Did you take the sign into account?
   c. What is the *change* in the momentum of each ball?

   Clay ball: Δp_x = _____   Rubber ball: Δp_x = _____
   d. Which ball experiences a larger impulse during the collision? Explain.

   e. From Newton's third law, the impulse that the ball exerts on the post is equal in magnitude, although opposite in direction, to the impulse that the post exerts on the ball. Which ball exerts the larger impulse on the post?

   f. Don't change your answer to part a, but are you still happy with that answer? If not, how would you change your answer? Why?

9.2 Momentum and the Impulse-Momentum Theorem

9.3 Solving Impulse and Momentum Problems

6. For each of the following situations, use both words and pictures to
   - Describe what happens in the language of force, acceleration, and action/reaction.
   - Describe what happens in the language of impulse and momentum.

a. A moving blob of clay hits a stationary bowling ball.

Force description:

Momentum description:

b. A falling rubber ball bounces off the floor.

Force description:

Momentum description:

c. Two equal masses are pushed apart by a compressed spring between them.

Force description:

Momentum description:

7. A small, light ball S and a large, heavy ball L move toward each other, collide, and bounce apart.

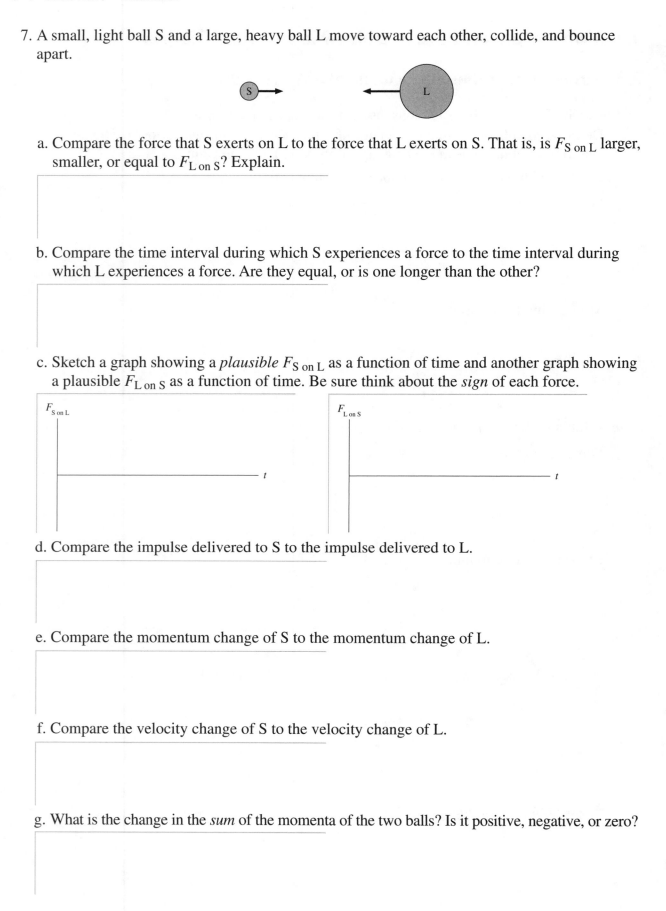

a. Compare the force that S exerts on L to the force that L exerts on S. That is, is $F_{S \text{ on } L}$ larger, smaller, or equal to $F_{L \text{ on } S}$? Explain.

b. Compare the time interval during which S experiences a force to the time interval during which L experiences a force. Are they equal, or is one longer than the other?

c. Sketch a graph showing a *plausible* $F_{S \text{ on } L}$ as a function of time and another graph showing a plausible $F_{L \text{ on } S}$ as a function of time. Be sure think about the *sign* of each force.

$F_{S \text{ on } L}$ t

$F_{L \text{ on } S}$ t

d. Compare the impulse delivered to S to the impulse delivered to L.

e. Compare the momentum change of S to the momentum change of L.

f. Compare the velocity change of S to the velocity change of L.

g. What is the change in the *sum* of the momenta of the two balls? Is it positive, negative, or zero?

Exercises 8–11: Prepare a pictorial representation for these problems, but *do not* solve them.
- Draw pictures of "before" and "after."
- Define symbols relevant to the problem.
- List known information, and identify the desired unknown.

8. A 50 kg archer, standing on frictionless ice, shoots a 100 g arrow at a speed of 100 m/s. What is the recoil speed of the archer?

9. The parking brake on a 2000 kg Cadillac has failed, and it is rolling slowly, at 1 mph, toward a group of small innocent children. As you see the situation, you realize there is just time for you to drive your 1000 kg Volkswagen side-on into the Cadillac and thus save the children. With what speed should you impact the Cadillac to bring it to a halt?

10. Dan is gliding on his skateboard at 4 m/s. He suddenly jumps backward off the skateboard, kicking the skateboard forward at 8 m/s. How fast is Dan going as his feet hit the ground? Dan's mass is 50 kg and the skateboard's mass is 5 kg.

11. While riding bumper cars at the fair, Bob's car collides directly with the back of Joe's car while both cars are moving to the right. Before the collision, Joe's car was traveling at 1.8 m/s and Bob's at 2.0 m/s. The combined mass of Joe and his car is only 80 kg, and the combined mass of Bob and his car is 100 kg. Immediately after the collision, Joe's car moves right at 2.0 m/s. How fast and in which direction does Bob's car move?

12. Identical blocks A and B are pushed to the right continuously by identical constant forces from the start to the finish shown below. At the starting line, block A is initially moving to the right, but block B is stationary. Which block undergoes a larger change in its momentum? Explain.

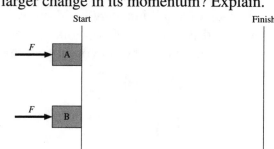

9.4 Conservation of Momentum

13. As you release a ball, it falls—gaining speed and momentum. Is momentum conserved?

    a. Answer this question from the perspective of choosing the ball alone as the system.

    b. Answer this question from the perspective of choosing ball+earth as the system.

14. Is it possible for momentum to be conserved if the system is NOT isolated? Explain.

15. Consider the interaction of two objects along a line. Show that the following three equations are equivalent if the system is isolated and the objects remain intact:

$$(p_f - p_i)_1 = -(p_f - p_i)_2$$

$$(p_i)_1 + (p_i)_2 = (p_f)_1 + (p_f)_2$$

$$F_{1 \text{ on } 2} = m_2 \frac{\Delta v_2}{\Delta t} = -F_{2 \text{ on } 1} = -m_1 \frac{\Delta v_1}{\Delta t}$$

9.5 Inelastic Collisions

16. For each of the following graphs a–e:
    - Indicate whether the graph could represent an inelastic collision between two objects A (solid line) and B (dashed line) at time t_0 and whether momentum could be conserved in each case. (Note that graph a shows position-versus-time and the remaining graphs show velocity-versus-time.)
    - What can you infer about the relative masses of A and B for those cases in which momentum is conserved?

a.

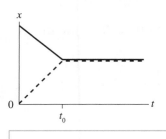

d.

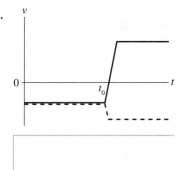

b.

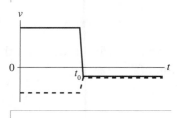

e.

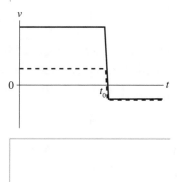

c.

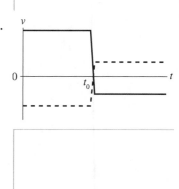

9.6 Momentum and Collisions in Two Dimensions

17. An object initially at rest explodes into three fragments.
    The momentum vectors of two of the fragments are shown.
    Draw the momentum vector $\vec{p}_3$ of the third fragment.

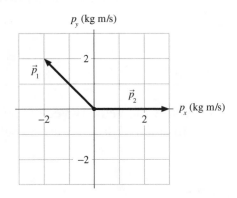

18. An object initially at rest explodes into three fragments.
    The momentum vectors of two of the fragments are shown.
    Draw the momentum vector $\vec{p}_3$ of the third fragment.

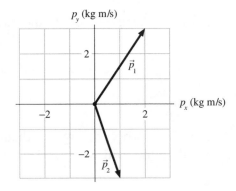

19. A 500 g ball traveling to the right at 4.0 m/s collides with
    and bounces off another ball. The figure shows the
    momentum vector $\vec{p}_1$ of the first ball after the collision.
    Draw the momentum vector $\vec{p}_2$ of the second ball.

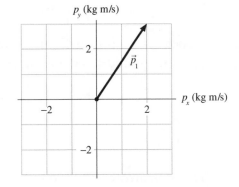

20. A 500 g ball traveling to the right at 4.0 m/s collides with
    and bounces off another ball. The figure shows the
    momentum vector $\vec{p}_1$ of the first ball after the collision.
    Draw the momentum vector $\vec{p}_2$ of the second ball.

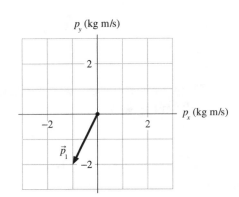

21. Consider a ball thrown at an angle against a wall with velocity vectors before and after the collision as shown at right from above.

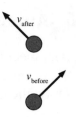

a. Redraw the velocity vectors below to determine the *change* in velocity of the ball due to the collision and draw an arrow below to indicate that change.

b. Draw an arrow below to indicate the direction of the impulse on the ball by the wall.

c. Draw an arrow below to indicate the relative magnitude and direction of the impulse by the ball on the wall.

d. Repeat part a for the figure shown at right. (The velocity vectors are the same magnitude as in part a.)

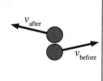

e. Draw an arrow below to indicate the direction of the impulse on the ball by the wall.

f. Draw an arrow below to indicate the relative magnitude and direction of the impulse by the ball on the wall.

9.7 Angular Momentum

22. A hoop of mass m and radius r is set spinning at angular speed ω about its center axis as shown from above.

    a. What would be its angular speed if its mass is suddenly doubled without changing its radius? Explain.

    b. What would be its angular speed if its radius is suddenly halved without changing its mass? Explain.

    c. What would be its angular speed if a force F is exerted at the outside surface of the cylinder directly toward the axis of rotation for a time Δt? Explain.

10 Energy and Work

10.1 A "Natural Money" Called Energy

1. One month, John has income of $3000, expenses of $2500, and he sells $300 of stocks.

   a. Can you determine John's liquid assets L at the end of the month? If so, what is L? If not, why not?

   b. Can you determine the amount by which John's liquid assets *changed* during the month? If so, what is ΔL?

2. John begins the month with $2000 of liquid assets and $5000 of savings. His financial activity for the month is as follows:

Day of Month	Activity
1	Receives a $3000 paycheck; deposits it in checking
3	Spends $500
8	Buys a $1000 savings bond
10	Pays bills totaling $1000
15	Receives a $100 birthday present from Grandma
23	Sells $1500 of stock
28	Buys a $1200 bicycle

   a. What are John's liquid assets and saved assets at the end of the month?

   b. Show that John's monetary relationship $\Delta W = I - E$ is satisfied.

10.2 The Basic Energy Model

3. What are the two primary processes by which energy can be transferred from the environment to a system?

4. Identify the energy transformations in each of the following processes (e.g., $K \rightarrow U_g \rightarrow E_{th}$)

   a. A ball is dropped from atop a tall building.

   b. A helicopter rises from the ground at constant speed.

   c. An arrow is shot from a bow and stops in the center of its target.

   d. A pole vaulter runs, plants his pole, and vaults up over the bar.

5. The kinetic energy of a system decreases and its potential energy is unchanged. What is doing work on what? That is, does the environment do work on the system, or does the system do work on the environment? Explain.

10.3 The Law of Conservation of Energy

6. Identify an appropriate system for applying conservation of energy to each of the following:

a. A spring is used to launch a ball into the air.

System:

b. A spring is used to push a car on an air track.

System:

c. A spring is used to slide a block across a table where it stops.

System:

d. A car moving on an air track collides with a spring and rebounds at essentially the same speed with which it hit the spring.

System:

7. What is meant by an *isolated system*?

8. a. A process occurs in which a system's potential energy decreases while the environment does work on the system. Does the system's kinetic energy increase, decrease, or stay the same? Or is there not enough information to tell? Explain.

b. A process occurs in which a system's potential energy increases while the environment does work on the system. Does the system's kinetic energy increase, decrease, or stay the same? Or is there not enough information to tell? Explain.

10.4 **Work**

9. For each situation described below:
   - Draw a before-and-after diagram, similar to Figures 10.8 and 10.11 in the textbook.
   - Identify *all* forces acting on the particle.
   - Determine if the work done by each of these forces is positive (+), negative (−), or zero (0). Make a little table beside the figure showing *every* force and the sign of its work.

   a. An elevator moves upward.

   b. An elevator moves downward.

   c. You push a box across a rough floor.

d. You slide down a steep hill.

e. A ball is thrown straight up. Consider the ball from one microsecond after it leaves your hand until the highest point of its trajectory.

f. A car turns a corner at constant speed.

10. A 0.2 kg plastic cart and a 20 kg lead cart both roll without friction on a horizontal surface. Equal forces are used to push both carts forward a distance of 1 m, starting from rest. After traveling 1 m, is the kinetic energy of the plastic cart greater than, less than, or equal to the kinetic energy of the lead cart? Explain.

11. An object experiences a force while undergoing the displacement shown. Is the work done positive (+), negative (−), or zero (0)?

a.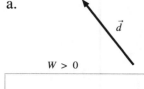

Sign = _____

b.

Sign = _____

c.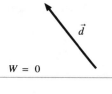

Sign = _____

d.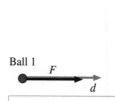

Sign = _____

e.

Sign = _____

f.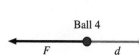

Sign = _____

12. Each of the diagrams below shows a displacement vector for an object. Draw and label a force vector that will do work on the object with the sign indicated.

a.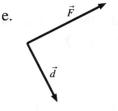

$W > 0$

b.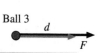

$W < 0$

c.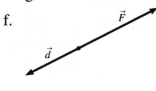

$W = 0$

13. Five balls with equal initial kinetic energies experience different forces acting over different displacements, as shown by the arrows. Rank in order, from largest to smallest, the kinetic energies of the balls after being acted on by these forces.

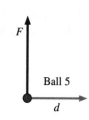

Order:

Explanation:

10.5 Kinetic Energy

14. Can kinetic energy ever be negative? _____

    Give a plausible *reason* for your answer without making use of any formulas.

15. a. If a particle's speed increases by a factor of three, by what factor does its kinetic energy change?

    b. Particle A has half the mass and eight times the kinetic energy of particle B. What is the speed ratio v_A/v_B?

    c. If a rotating skater triples her rate of rotation by decreasing her moment of inertia by 1/3, by what factor does her rotational kinetic energy change?

16. On the axes below, draw graphs of the kinetic energy of

    a. A 1000 kg car that uniformly accelerates from 0 to 20 m/s in 20 s.
    b. A 1000 kg car moving at 20 m/s that brakes to a halt with uniform deceleration in 4 s.
    c. A 1000 kg car that drives once around a 40-m-diameter circle at a speed of 20 m/s.

    Calculate K at several times, plot the points, and draw a smooth curve between them.

a.
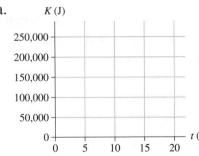

b.
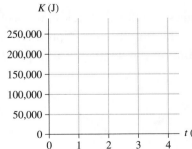

c.
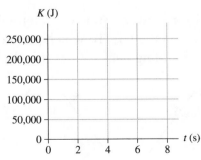

10.6 Potential Energy

17. Below we see a 1 kg object that is initially 1 m above the ground and rises to a height of 2 m. Anjay and Brittany each measure its position but use a different coordinate system to do so. Fill in the table to show the initial and final gravitational potential energies and ΔU as measured by Anjay and Brittany.

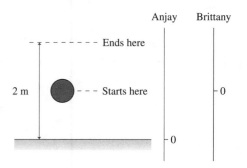

	U_i	U_f	ΔU
Anjay			
Brittany			

18. Three balls of equal mass are fired simultaneously with *equal* speeds from the same height above the ground. Ball 1 is fired straight up, ball 2 is fired straight down, and ball 3 is fired horizontally. Rank in order, from largest to smallest, their speeds v_1, v_2, and v_3 as they hit the ground.

Order:

Explanation:

19. Below are shown three frictionless tracks. A block is released from rest at the position shown on the left. To which point does the block make it on the right before reversing direction and sliding back? Point B is the same height as the starting position.

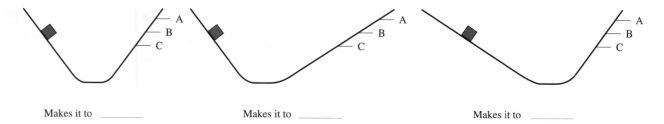

Makes it to _____   Makes it to _____   Makes it to _____

20. Two balls are released from just below the rim of two identical bowls. Ball 1 rolls down without slipping while ball 2 slides down without friction. Which ball will reach the higher point on the other side before reversing direction? Explain.

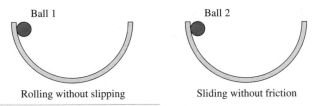

Rolling without slipping   Sliding without friction

21. A heavy object is released from rest at position 1 above a spring. It falls and contacts the spring at position 2. The spring achieves maxiumum compression at position 3. Fill in the table below to indicate whether each of the quantities are +, −, or 0 during the intervals 1→2, 2→3, and 1→3.

	1→2	2→3	1→3
ΔK			
ΔU_g			
ΔU_s			

22. Rank in order, from most to least, the amount of elastic potential energy $(U_s)_1$ to $(U_s)_4$ stored in each of these springs.

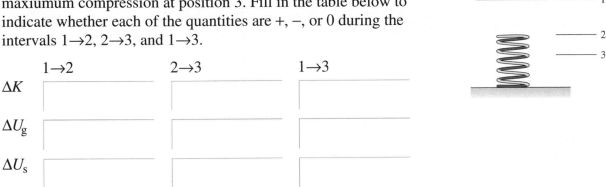

1 k Compressed d 2 k Stretched d 3 $2k$ Stretched d 4 k Stretched $2d$

Order:

Explanation:

10.7 Thermal Energy

23. A car traveling at 60 mph slams on its brakes and skids to a halt. What happened to the kinetic energy the car had just before stopping?

24. What energy transformations occur as a skier glides down a gentle slope at constant speed?

25. Give a *specific* example of a situation in which:

    a. $W \rightarrow K$ with $\Delta U = 0$ and $\Delta E_{th} = 0$.

    b. $W \rightarrow U$ with $\Delta K = 0$ and $\Delta E_{th} = 0$.

    c. $K \rightarrow U$ with $W = 0$ and $\Delta E_{th} = 0$.

    d. $W \rightarrow E_{th}$ with $\Delta K = 0$ and $\Delta U = 0$.

    e. $U \rightarrow E_{th}$ with $\Delta K = 0$ and $W = 0$.

10.8 Further Examples of Conservation of Energy

26. If a solid disk and a circular hoop of the same mass and radius are released from rest at the top of a ramp and allowed to roll to the bottom, the disk will get to the bottom first. *Without referring to equations*, explain why this is so.

10.9 Energy in Collisions

27. Ball 1 with an initial speed of 14 m/s has a perfectly elastic collision with ball 2 that is initially at rest. Afterward, the speed of ball 2 is 21 m/s.

    a. What will be the speed of ball 2 if the initial speed of ball 1 is doubled?

    b. What will be the speed of ball 2 if the mass of ball 1 is doubled?

28. You can dive into a swimming pool of water from a high diving board without being hurt, but to dive into an empty pool from a much lower distance might be fatal? Why the difference?

29. Consider a perfectly elastic collision in which a moving ball 1 strikes an initially stationary ball 2. Is ball 1 more likely to recoil backwards if it is moving very fast (large forward momentum) or moving slowly (small forward momentum) assuming that ball 2 is identical in each case? Or does the speed of ball 1 matter? Explain.

30. Consider a perfectly elastic collision in which a moving ball 1 strikes a initially stationary ball 2.
   a. Under what circumstances, if any, will ball 1 come to a stop?

   b. Under what circumstances, if any, will ball 1 recoil backwards?

   c. Under what circumstances, if any, will ball 1 continue moving forward?

   d. Is it possible for ball 1 to move forward or backwards at a greater speed than its speed just before the collision?

10.10 Power

31. a. If you push an object 10 m with a 10 N force in the direction of motion, how much work do you do on it?

   b. How much power must you provide to push the object in 1 s? In 10 s? In 0.1 s?

11 Using Energy

11.1 Transforming Energy

1. a. Given that efficiency is defined as e = (what you get)/(what you had to pay), how might one define the concept of "inefficiency"?

   b. Write a mathematical relationship to relate your definition of inefficiency, which you can denote as i, to the definition of efficiency e.

2. A device uses 500 J of chemical energy to generate 100 J of electric energy.
   a. What is the efficiency of this device?

   b. A second device is twice as efficient. How much electric energy can it generate from the same 500 J of chemical energy?

3. Is it possible for the efficiency of some device or machine to be greater than 1? Either give an example where the efficiency is greater than 1 or explain why it's not possible.

4. A small battery-operated car moves at constant velocity across the level floor of a classroom.
   a. Is work being done on the car? If so, by what force or forces?

   b. Is energy being used? If so, what is the source of this energy?

11.2 Energy in the Body: Energy Inputs

5. Can you run further by burning an ounce of fat or an ounce of carbohydrates? Explain.

6. Which has more energy content, three apples or one slice of apple pie? Why?

11.3 Energy in the Body: Energy Outputs

7. The energy used by a human while running is approximately proportional to the speed. If we expressed the energy used by a runner in terms of, say, miles per candy bar, who gets better mileage per candy bar, a runner who completes a marathon in $2\frac{1}{2}$ hours or one who completes the same race in 4 hours? Explain.

8. Suppose your body burns 10 Cal climbing stairs. Which of the following will allow you to burn 20 Cal?
   a. Climb twice as high at the same speed.
   b. Climb the original stairs twice as fast.
   c. Either a or b.
   d. Neither a nor b.

   Explain.

11.4 Thermal Energy and Temperature

9. Rank in order, from highest to lowest, the temperatures $T_1 = 0$ K, $T_2 = 0°C$, and $T_3 = 0°F$.

10. "Room temperature" is often considered to be 68°F. What is room temperature in °C and in K?

11. a. What is the average translational kinetic energy of a gas at absolute zero?

   b. Can an atom have negative kinetic energy? Explain.

   c. Based on your answers to parts a and b, what is the translational kinetic energy of *every* atom in the gas?

   d. Would it be physically possible for the thermal energy of a gas to be less than its thermal energy at absolute zero? Explain.

   e. Is it possible to have a temperature less than absolute zero? Explain.

12. The quantity y is proportional to the square root of x, and $y = 12$ when $x = 16$.

  a. Write an equation to represent this square root relationship for all x and y.

  b. Find y if $x = 9$.

  c. Find x if $y = 6$.

  d. By what factor must x change for the value of y to double?

  e. Compare your equation in part a to the equation from your text relating v_{rms} and T, $v_{rms} = \sqrt{\dfrac{3k_{B}T}{m}}$. Which quantity assumes the role of x? Which quantity assumes the role of y? What is the constant of proportionality relating v_{rms} and T?

  f. Rewrite your equation in part a as a quadratic relationship with x in terms of y^2.

  g. Rewrite $v_{rms} = \sqrt{\dfrac{3k_{B}T}{m}}$ as a quadratic relationship between v_{rms} and T.

13. If you double the temperature of a gas:

    a. Does the average translational kinetic energy per atom change? If so, by what factor?

    b. Does the root-mean-square speed of the atoms change? If so, by what factor?

14. Suppose you could suddenly increase the speed of every atom in a gas by a factor of 2.

    a. Would the rms speed of the atoms increase by a factor of $(2)^{1/2}$, 2, or 2^2? Explain.

    b. Would the temperature of the gas increase by a factor of $(2)^{1/2}$, 2, or 2^2? Explain.

15. Lithium vapor, which is produced by heating lithium to the relatively low boiling point of 1340°C, forms a gas of Li_2 molecules. Each molecule has a molecular mass of 14 u. The molecules in nitrogen gas (N_2) have a molecular mass of 28 u. If the Li_2 and N_2 gases are at the same temperature, which of the following is true?

    a. v_{rms} of $N_2 = 2.00 \times v_{rms}$ of Li_2.
    b. v_{rms} of $N_2 = 1.41 \times v_{rms}$ of Li_2.
    c. v_{rms} of $N_2 = v_{rms}$ of Li_2.
    d. v_{rms} of $N_2 = 0.71 \times v_{rms}$ of Li_2.
    e. v_{rms} of $N_2 = 0.50 \times v_{rms}$ of Li_2.

    Explain.

11.5 Heat and the First Law of Thermodynamics

16. Do each of the following describe a property of a system, an interaction of a system with its environment, or both? Explain.

a. Temperature:

b. Heat:

c. Thermal energy:

17. For each of the following processes:

a. Is the value of the work W, the heat Q, and the change of thermal energy ΔE_{th} positive (+), negative (−) or zero (0)?

b. Does the temperature increase (+), decrease (−), or not change (0)?

	W	Q	ΔE_{th}	ΔT
• You hit a nail with a hammer.				
• You hold a nail over a Bunsen burner.				
• High-pressure steam spins a turbine.				
• Steam contacts a cold surface and condenses.				
• A moving crate slides to a halt on a rough surface.				

11.6 Heat Engines

18. Rank in order, from largest to smallest, the efficiencies e_1 to e_4 of these heat engines.

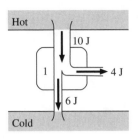
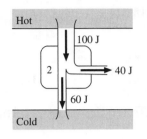
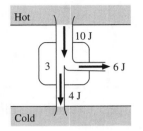
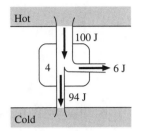

Order:

Explanation:

19. For each engine shown,
    a. Supply the missing value.
    b. Determine the efficiency.

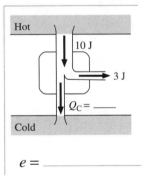
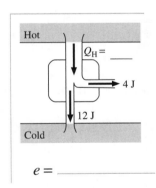
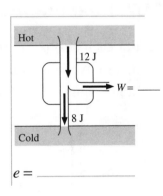

$e =$ _____ $e =$ _____ $e =$ _____

20. Efficiency is a dimensionless quantity, so why is it necessary to measure temperatures in kelvin rather than °C or °F to determine efficiency when using the equation $e = 1 - \dfrac{T_C}{T_H}$?

21. Four heat engines with maximum efficiency (Carnot engines) operate with the hot and cold reservoir temperatures shown in the table.

Engine	T_C (K)	T_H (K)
1	300	600
2	200	400
3	200	600
4	300	400

Rank in order, from largest to smallest, the efficiencies e_1 to e_4 of these engines.

Order:

Explanation:

11.7 Heat Pumps

22. For each heat pump shown,
    a. Supply the missing value.
    b. Determine the coefficient of performance if the heat pump is used for cooling.

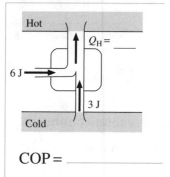

COP = _____

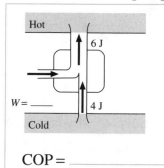

COP = _____

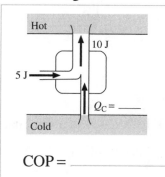

COP = _____

23. Does a refrigerator do work in order to cool the interior? Explain.

11.8 Entropy and the Second Law of Thermodynamics

11.9 Systems, Energy and Entropy

24. Do each of the following represent a possible heat engine or heat pump? If not, what is wrong?

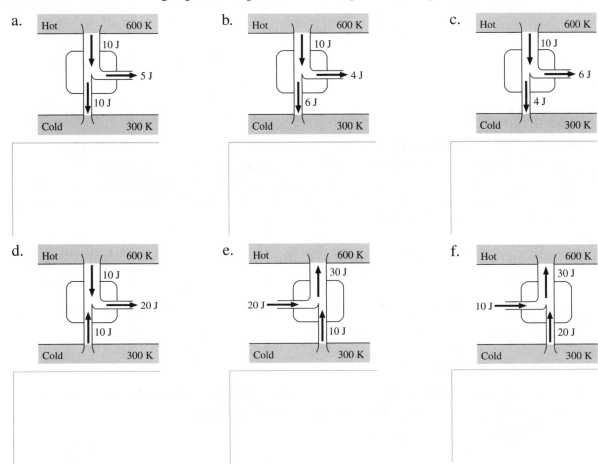

25. If you place a jar of perfume in the center of a room and remove the stopper, you will soon be able to smell the perfume throughout the room. If you wait long enough, will all the perfume molecules ever be back in the jar at the same time? Why or why not?

26. Suppose you place an ice cube in a cup of room-temperature water and then seal them in a well-insulated container. No energy can enter or leave the container.

    a. If you open the container an hour later, which do you expect to find: a cup of water, slightly cooler than room temperature, or a large ice cube and some 100°C steam?

    b. Finding a large ice cube and some 100°C steam would not violate the first law of thermodynamics. $W = 0$ J and $Q = 0$ J, because the container is sealed, and $\Delta E_{th} = 0$ J because the increase in thermal energy of the water molecules that have become steam is offset by the decrease in water molecules that have turned to ice. Energy is conserved, yet we never see a process like this. Why not?

27. Are each of the following processes reversible or irreversible? Would the second law of thermodynamics be violated by any of the processes? Explain.

    a. A freshly baked pie cools on a window sill.

    b. A person bounces up from a firemen's trampoline onto the ledge of a burning building.

    c. A neatly raked pile of leaves is scattered by the wind.

12 Thermal Properties of Matter

12.1 The Atomic Model of Matter

1. The cube shown is made up of eight identical cubes, set $2 \times 2 \times 2$.

   a. In the space at right, draw a block that has twice the total volume but made out of cubes of the same size.

   b. How many cubes are needed to construct the larger block? _____

   c. How many cubes are needed to construct a solid block that is twice as large in each dimension $(4 \times 4 \times 4)$? _____

   b. By what factor is the volume of the block in part c greater than the volume of the original $2 \times 2 \times 2$ block? _____

2. a. Solids and liquids resist being compressed. They are not totally incompressible, but it takes large forces to compress them even slightly. If it is true that matter consists of atoms, what can you infer about the microscopic nature of solids and liquids from their incompressibility?

   b. Solids also resist being pulled apart. You can break a metal or glass rod by pulling the ends in opposite directions, but it takes a large force to do so. What can you infer from this observation about the properties of atoms?

12.2 Thermal Expansion

3. The figure shows a thin square solid object with a circular hole in the center. The hole has a diameter that is $\frac{1}{2}$ the length of each side of the square. Redraw the object in the space at right after the object has undergone an expansion of its size by 50% in each linear dimension.

4. a. Solids are characterized by both a coefficient of linear expansion and a coefficient of volume expansion. Examine Table 12.3 and describe the relationship between these two coefficients.

   b. Liquids, in contrast with solids, have only a coefficient of volume expansion. Why don't liquids have a coefficient of linear expansion?

   c. Consider two equal volumes of liquid, one held in a tall narrow container and the other in a short wide container.
   i. Does the linear height of the liquid in each case change by the same absolute amount for a given temperature change?

   ii. Does it change by the same percentage amount? Explain.

12.3 Pressure and the Kinetic Theory of an Ideal Gas

5. a. If you double the absolute pressure of an ideal gas contained in a fixed volume at constant temperature, by how much does the gauge pressure change?

   b. If you double the gauge pressure of an ideal gas contained in a fixed volume at constant temperature, by how much does the absolute pressure change?

6. Gases, in contrast with solids and liquids, are very compressible. What can you infer from this observation about the microscopic nature of gases?

7. a. Can you think of any everyday experiences or observations that would suggest that the molecules of a gas are in constant, *random* motion? (Note: The existence of "wind" is *not* such an observation. Wind implies that the gas as a whole can move, but it doesn't tell you anything about the motions of the individual molecules in the gas.)

b. If the molecules are moving randomly and colliding with each other, do you expect there to be a very wide range of molecular speeds? Or do you expect that most of the molecules will have very similar speeds? Explain.

8. Consider an ideal gas contained in a confined volume. How would the pressure of the gas change if

a. the number of molecules of the gas were doubled, without changing the container or the temperature?

b. the volume of the container were doubled, without changing the number of molecules or the temperature?

c. the temperature (in K) of the gas were doubled, without changing the number of molecules or the volume of the container?

d. the rms speed of the molecules were doubled, without changing the number of molecules or the volume of the container?

12.4 Ideal-Gas Processes

9. The graphs below show the initial state of a gas. Draw a *pV* diagram showing the following processes:

    a. A constant-volume process that doubles the pressure.
    b. An isobaric process that doubles the temperature.
    c. An isothermal process that halves the volume.

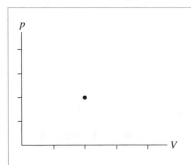 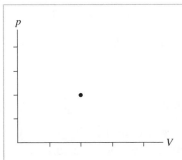 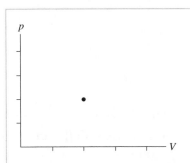

10. Interpret the *pV* diagrams shown below by
    a. Naming the process.
    b. Stating the *factors* by which *p*, *V*, and *T* change. (A fixed quantity changes by a factor of 1.)

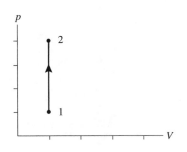 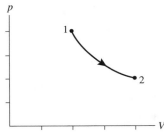 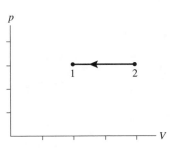

Process _____	Process _____	Process _____
p changes by _____	*p* changes by _____	*p* changes by _____
V changes by _____	*V* changes by _____	*V* changes by _____
T changes by _____	*T* changes by _____	*T* changes by _____

11. Starting from the initial state shown, draw a *pV* diagram for the three-step process:
    i. A constant-volume process that halves the temperature.
    ii. An isothermal process that halves the pressure, then
    iii. An isobaric process that doubles the volume.
    Label each of the stages on your diagram.

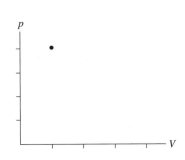

12. How much work is done on the gas in each of the following processes?

a.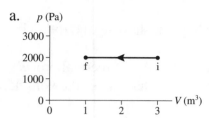

b.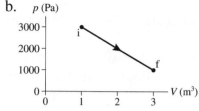

c.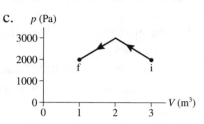

W = _____ W = _____ W = _____

13. The figure below shows a process in which a gas is compressed from 300 cm^3 to 100 cm^3.

   a. Use the middle set of axes to draw the pV diagram of a process that starts from initial state i, compresses the gas to 100 cm^3, and does the same amount of work on the gas as the process on the left.

   b. Is there a constant-volume process that does the same amount of work on the gas as the process on the left? If so, show it on the axes on the right. If not, use the blank space of the axes to explain why.

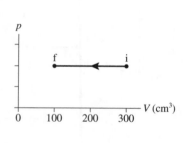 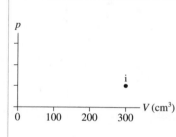 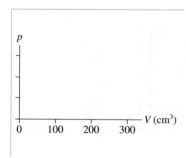

14. The figure shows a process in which work is done to compress a gas.

   a. Draw and label a process A that starts and ends at the same points but does more work on the gas.

   b. Draw and label a process B that starts and ends at the same points but does less work on the gas.

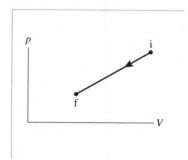

15. The figure shows an adiabatic process.

a. Is the final temperature higher than, lower than, or equal to the initial temperature? Explain.

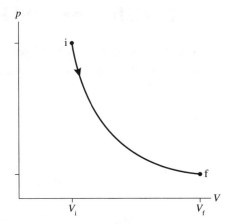

b. Draw *and label* the T_i and T_f isotherms on the figure.

c. Is the work done on the gas positive or negative? Explain.

d. Show *on the figure* how you would determine the amount of work done.

e. Is any heat energy added to or removed from the system in this process? Explain.

f. *Why* does the gas temperature change?

16. Starting from the point shown, draw a *pV* diagram for the following processes.

a. An isobaric process in which work is done *by* the system.

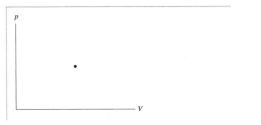

b. An adiabatic process in which work is done *on* the system.

c. An isothermal process in which heat is *added to* the system.

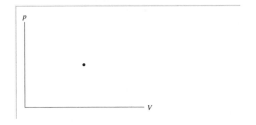

d. A constant-volume process in which heat is *removed from* the system.

12.5 Specific Heat and Heat of Transformation

17. 400 Joules of energy are used to change the temperature of a 0.250 kg sample by 5 K.
    a. What does the ratio $\frac{400}{5}$ mean in this context?

    b. What does the ratio $\frac{400}{0.250}$ mean in this context?

    c. What does the ratio $\frac{400}{0.250 \times 5}$ mean in this context?

18. Explain in your own words the meaning of the statement that the specific heat of a liquid is 4000 J/kg·K.

19. On the phase diagram:

    a. Draw *and label* a line to show a process in which the substance boils at constant pressure. Be sure to include an arrowhead on your line to show the direction of the process.

    b. Draw *and label* a line to show a process in which the substance freezes at constant temperature.

    c. Draw *and label* a line to show a process in which the substance sublimates (changes directly from solid to gas) at constant pressure.

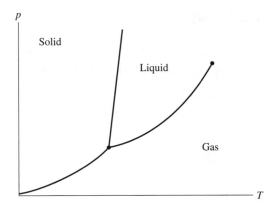

20. The figure shows the phase diagram of water. Answer the following questions by referring to the phase diagram and, perhaps, drawing lines on the phase diagram.

    a. What happens to the boiling-point temperature of water as you go to higher and higher elevations in the mountains?

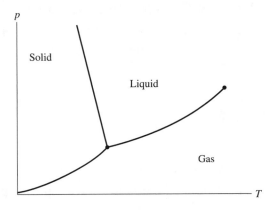

    b. Suppose you place a beaker of liquid water at 20°C in a vacuum chamber and then steadily reduce the pressure. What, if anything, happens to the water?

12.6 Calorimetry

21. A beaker of water at 80.0°C is placed in the center of a large, well-insulated room whose air temperature is 20.0°C. Is the final temperature of the water:

    i.   20.0°C.
    ii. Slightly above 20.0°C.
    iii. 50.0°C.
    iv. Slightly less than 80.0°C.
    v.   80.0°C.

    Explain.

22. Cold water is poured into a hot metal container.

    a. What physical quantities can you *measure* that tell you that the metal and water are somehow changing?

    b. What is the condition for equilibrium, after which no additional changes take place?

    c. Use the concept of energy to describe how the metal and the water interact.

    d. Is your description in part c something that you can *observe* happening? Or is it an *inference* based on the measurements you specified in part a?

12.7 Thermal Properties of Gases

23. Describe *why* the molar specific heat at constant pressure is larger than the molar specific heat at constant volume.

24. The figure on the left shows a thermodynamic process in which a gas expands from 100 cm^3 to 300 cm^3. On the right, draw the pV diagram of a process that starts from state i, expands to 300 cm^3, and does the same amount of work as the process on the left.

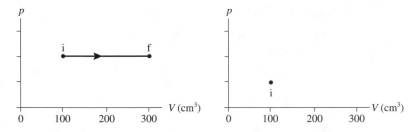

25. For each of these processes, is work done *by* the system ($W < 0$, $W_s > 0$), *on* the system ($W > 0$, $W_s < 0$), or is *no* work done?

a.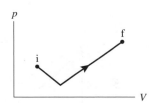

b.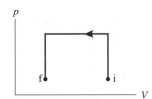

c.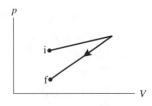

Work is _____   Work is _____   Work is _____

26. Rank in order, from largest to smallest, the amount of work W_1 to W_4 done by the gas in each of these cycles.

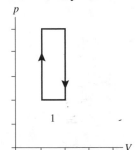   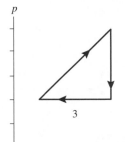   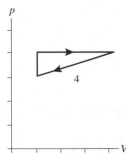

Order:

Explanation:

27. The figure uses a series of pictures to illustrate a thermodynamic cycle.

Stage 1 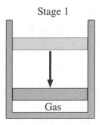 Stage 2 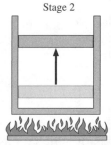 Pin 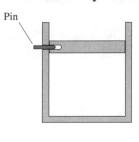 Stage 3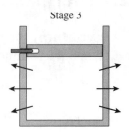

The gas is compressed rapidly from V_1 to V_2.

The gas is heated at constant temperature until the volume returns to V_1.

The flame is turned off and the piston is locked in place.

The gas cools until the initial pressure p_1 is restored.

a. Show the cycle as a pV diagram. Label the three stages.

b. What is the energy transformation during each stage of the process? (For example, a stage in which work energy is transformed into heat energy could be represented as $W \rightarrow Q$.)

Stage 1: _____

Stage 2: _____

Stage 3: _____

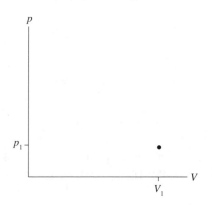

28. The figure shows the pV diagram of a heat engine.
   a. During which stages is heat added to the gas? _____
   b. During which is heat removed from the gas? _____
   c. During which stages is work done on the gas? _____
   d. During which is work done by the gas? _____
   e. Draw a series of pictures, similar to those of Exercise 10, to illustrate the stages of this cycle. Give a brief description of what happens during each stage.

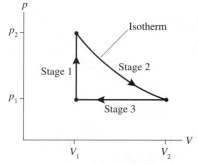

12.8 Heat Transfer

29. Hot water is poured into a cold container. Give a *microscopic* description of how these two systems interact until they reach thermal equilibrium.

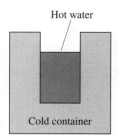

13 Fluids

13.1 Fluids and Density

1. An object has density ρ.

   a. Suppose each of the object's three dimensions is increased by a factor of 2 without changing the nature of the material of which the object is made. Will the density change? If so, by what factor? Explain.

   b. Suppose each of the object's three dimensions is increased by a factor of 2 without changing the object's mass. Will the density change? If so, by what factor? Explain.

2. A stone cutter cuts a section off of a marble slab that is one-eighth the weight of the original slab. How does the density of the cut section compare to the density of the original slab? Explain.

3. A cylinder contains 2 g of oxygen gas. A piston is used to compress the gas. After the gas has been compressed:

   a. Has the mass of the gas increased, decreased, or not changed? Explain.

   b. Has the density of the gas increased, decreased, or not changed? Explain.

13.2 Pressure

13.3 Measuring and Using Pressure

4. The four containers marked A through D shown are filled with water to equal depths. How do the pressures compare at the depths indicated by the dotted line? Explain.

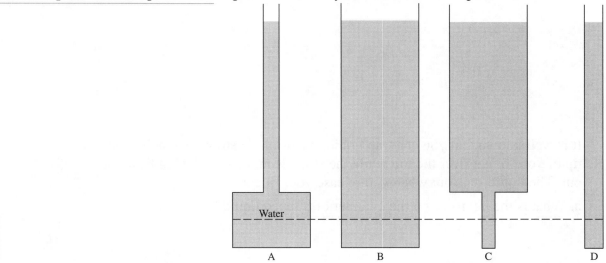

5. Write equations that would allow you to compare the pressures at the bottoms of the following containers of water. There is no need to solve your equations here. Container A contains water to a depth of 20 cm under a 10 cm² water-tight piston that is held down by a 111 N weight. Container B is the same shape as container A except that it is topped instead by a long tube filled to a height of 1.0 m with water and open to atmospheric pressure at the top. Container C is similar to container A except that it is sealed at the top with a gas chamber under a gauge pressure of 0.1 atmospheres.

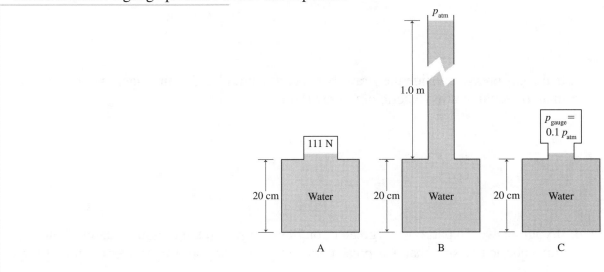

6. The figure shows a manometer comparable to the one in Figure 13.12, except that the height of the liquid on the left side is higher than that on the right. Is this situation possible? If not, why not? If so, what can you say about the gas pressure in the tank attached to the left side column?

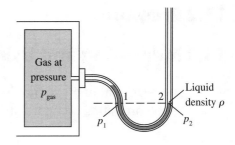

7. It is well known that you can trap liquid in a drinking straw by placing the tip of your finger over the top while the straw is in the liquid, then lifting it out. The liquid runs out when you release your finger.

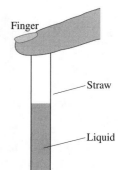

 a. What is the *net* force on the cylinder of trapped liquid?

 b. Draw a free-body diagram for the trapped liquid. Label each vector.

 c. Is the gas pressure inside the straw, between the liquid and your finger, greater than, less than, or equal to atmospheric pressure? Explain.

 d. If your answer to part c was "greater" or "less," how did the pressure change from the atmospheric pressure that was present when you placed your finger over the top of the straw?

13.4 Buoyancy

8. Three blocks of identical size, A, B, and C, are to be gently placed into a large tank of water. Block A has a density of 2 g/cm^3, block B has a density of 0.9 g/cm^3, and block C has a density of 0.5 g/cm^3.

a. Sketch the equilibrium locations for blocks in the water tank shown.

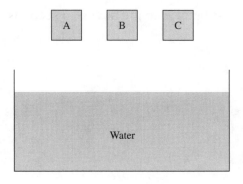

b. In the space below, draw a free-body diagram for each of blocks A, B, and C.

c. What would the equilibrium position in the water tank be for a fourth block of density 1.0 g/cm^3? Explain.

13.5 Fluids in Motion

13.6 Fluid Dynamics

9. A stream flows at constant depth through the channel shown in an overhead view. (A one-m^2 grid has been superposed on the channel to facilitate measurement.) The flow speed at point A is a constant 2 meters per second to the right.
   a. Shade in squares to represent the water that will flow past point A in the next two seconds.

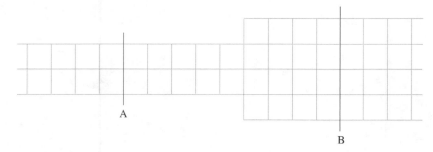

   b. The water then empties into the wider channel shown. Assuming a continuous flow of water, shade in squares to represent the water that will flow to the right past point B in the next two seconds. Explain.

10. Liquid flows through the pipe shown. An overhead view shows the varying width of the pipe at different locations. The fluid pressure in the pipe can be determined by the level of liquid in the three vertical pipes A, B, and C of the side view. The level of liquid in the pipe A is shown, but not in the pipes B and C. Draw an appropriate level of liquid in pipes B and C to indicate the relative pressures at those points. Explain.

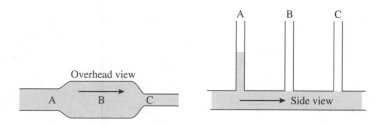

11. Streamlines are shown for flow through a stream.

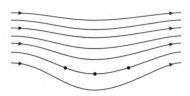

    a. Sketch force vectors on the stream lines at the points shown by dots.
    b. Where is the pressure the greatest in the region shown? Where is it the least? Explain.

12. In Chapter 10, we saw that for a system in which only mechanical work is involved, one can write the law of conservation of energy as $\Delta K + \Delta U_g = W$. Compare this equation to Bernoulli's equation. Is Bernoulli's equation a restatement of conservation of energy for fluids?

13.7 Viscosity and Poiseuille's Equation

13. Restate Poiseuille's equation $v_{avg} = \frac{R^2}{8\eta L}\Delta p$ in your own words, including a description of the property expressed by each symbol in the equation as well as the object to which that property is attributed. It is not sufficient, for example, to state that the left-hand side is the "average velocity" without stating the average velocity of what.

14. If the radius of a pipe filled with a flowing fluid is doubled, by how much must the length be changed so that the average velocity of a fluid flowing through the pipe will be the same under the same pressure difference across its ends?

14 Oscillations

14.1 Equilibrium and Oscillation

14.2 Linear Restoring Forces and Simple Harmonic Motion

1. On the axes below, sketch three cycles of the position-versus-time graph for:

    a. A particle undergoing simple harmonic motion.

    b. A particle undergoing periodic motion that is not simple harmonic motion.

2. Consider the particle whose motion is represented by the x-versus-t graph below.

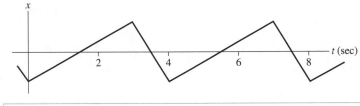

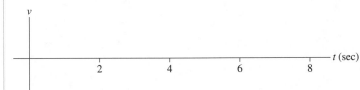

    a. Is this periodic motion? _____   b. Is this motion SHM? _____

    c. What is the period? _____   d. What is the frequency? _____

    e. You learned in Chapter 2 to relate velocity graphs to position graphs. Use that knowledge to draw the particle's velocity-versus-time graph on the axes provided.

3. A mass hung vertically on the spring shown in A stretches the spring by ΔL as shown in B, which is in its equilibrium position. Draw a free-body diagram for the same mass-spring system shown in locations B, C (in which the spring is stretched by $2\Delta L$), and D (in which the spring is stretched by $\Delta L/2$).

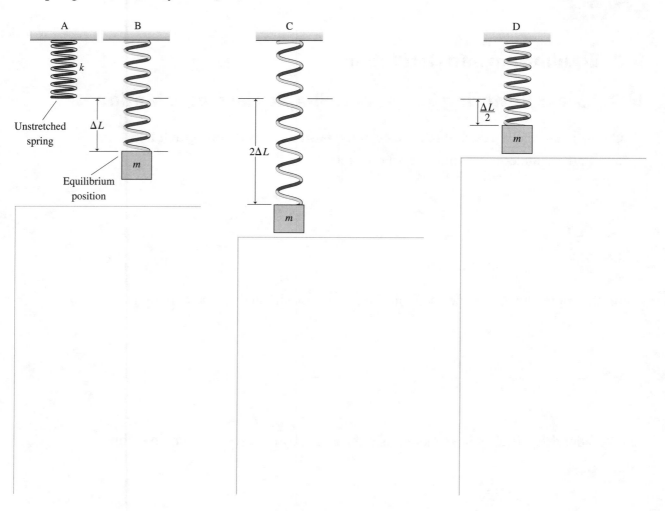

14.3 Describing Simple Harmonic Motion

4. The graph shown is the position-versus-time graph of an oscillating particle. It is constructed of *parabolic* segments that are joined at $x = 0$.

   a. Is this simple harmonic motion? Why or why not?

   b. Draw the corresponding velocity-versus-time graph.
      Hint: Recall that each parabolic segment corresponds to linearly changing velocity and constant positive or negative acceleration.

   c. Draw the corresponding acceleration-versus-time graph.

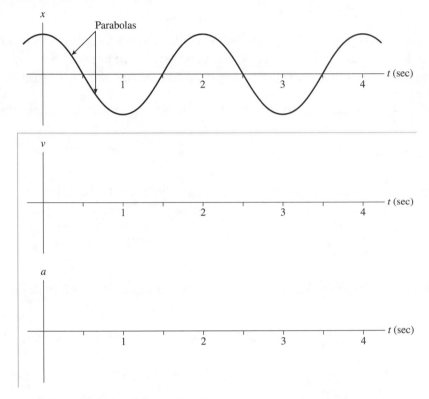

   d. At what times is the position a maximum? _____

      At those times, is the velocity a maximum, a minimum, or zero? _____

      At those times, is the acceleration a maximum, a minimum, or zero? _____

   e. At what times is the position a minimum (most negative)? _____

      At those times, is the velocity a maximum, a minimum, or zero? _____

      At those times, is the acceleration a maximum, a minimum, or zero? _____

   f. At what times is the velocity a maximum? _____

      At those times, where is the particle? _____

   g. Can you find a simple relationship between the *sign* of the position and the *sign* of the acceleration at the same instant of time? If so, what is it?

14-4 CHAPTER 14 · Oscillations

5. A particle goes around a circle 5 times at constant speed, taking a total of 2.5 seconds.

   a. Through what angle *in degrees* has the particle moved? _____

   b. Through what angle *in radians* has the particle moved? _____

   c. What is the particle's frequency f?

   d. Use your answer to part b to determine the particle's angular frequency ω.

   e. Does ω (in rad/s) $= 2\pi f$ (in Hz)? _____

6. The quantity $\sin\theta$ is familiar from trigonometry as the ratio of a right triangle's opposite side to its hypotenuse. In that case, it is usually equally convenient to express θ in either degrees or radians. By contrast, if a sinusoidal function is used to describe simple harmonic motion, such as $x = A\sin\left(\dfrac{2\pi t}{T}\right)$, the angle θ is itself a function of other quantities, namely the period T and the time t, where $\theta = \dfrac{2\pi t}{T}$.

   a. Why is it necessary to use radians as the argument for the sine function in $x = A\sin\left(\dfrac{2\pi t}{T}\right)$?

   b. How could the sine function be revised so that its argument is given in degrees?

   c. The function $x = A\sin\left(\dfrac{2\pi t}{T}\right)$ has the value 0 when $t = 0$ and has its first maximum value A when $t = T/4$. What is x at $t = T/8$?

   d. At what earliest time does $x = A/2$?

7. A particle moves counterclockwise around a circle at constant speed beginning at $\phi_0 = 0$. For each of the kinematic quantities shown, sketch two cycles of the component of the particle's motion.

a.
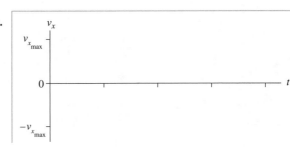

b.

c.
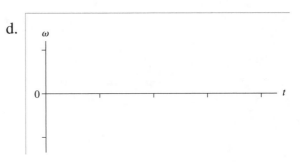

d.

8. A mass on a spring oscillates with period T, amplitude A, maximum speed v_{max}, and maximum acceleration a_{max}.

a. If T doubles without changing A,

i. how does v_{max} change, if at all?

ii. how does a_{max} change, if at all?

b. If A doubles without changing T,

i. how does v_{max} change, if at all?

ii. how does a_{max} change, if at all?

14.4 Energy in Simple Harmonic Motion

9. The figure shows the potential-energy diagram and the total energy line of a particle oscillating on a spring.

    a. What is the spring's equilibrium length?

    b. Where are the turning points of the motion? Explain how you identify them.

    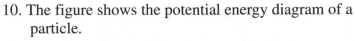

    c. What is the particle's maximum kinetic energy?

    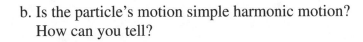

    d. Draw a graph of the particle's kinetic energy as a function of position.

    e. What will be the turning points if the particle's total energy is doubled?

    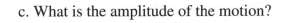

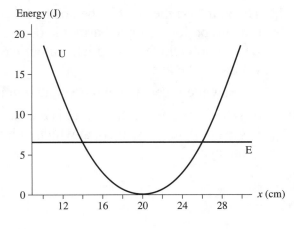

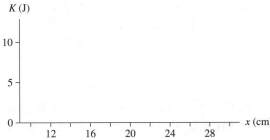

10. The figure shows the potential energy diagram of a particle.

    a. Is the particle's motion periodic? How can you tell?

    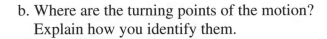

    b. Is the particle's motion simple harmonic motion? How can you tell?

    c. What is the amplitude of the motion?

    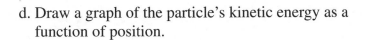

    d. Draw a graph of the particle's kinetic energy as a function of position.

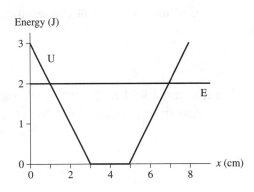

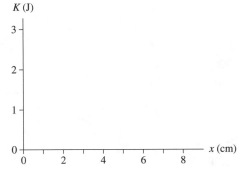

11. The graph on the right is the position-versus-time graph for a simple harmonic oscillator. Be sure to include appropriate numerical values on the axes.

    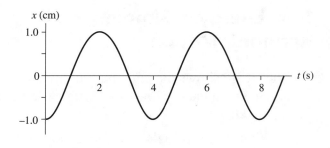

    a. Draw the v-versus-t and a-versus-t graphs.

    b. When x is greater than zero, is a ever greater than zero? If so, at which points in the cycle?

    c. When x is less than zero, is a ever less than zero? If so, at which points in the cycle?

    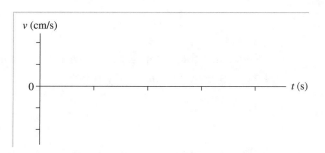

    d. Can you make a general conclusion about the relationship between the sign of x and the sign of a?

    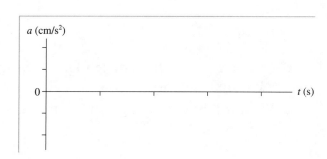

    e. When x is greater than zero, is v ever greater than zero? If so, how is the oscillator moving at those times?

12. Equation 14.24 in the textbook states that $\frac{1}{2}kA^2 = \frac{1}{2}mv_{max}^2$. What does this equation mean? Is there a time at which both the speed and displacement are at maximum? Explain.

13. The top graph shows the position versus time for a mass oscillating on a spring. On the axes below, sketch the position-versus-time graph for this block for the following situations:

Note: The changes described in each part refer back to the original oscillation, not to the oscillation of the previous part of the question. Assume that all other parameters remain constant. Use the same horizontal and vertical scales as the original oscillation graph.

a. The amplitude and the frequency are doubled.

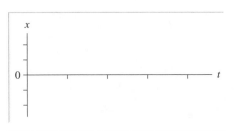

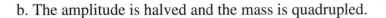

b. The amplitude is halved and the mass is quadrupled.

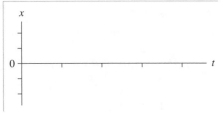

c. The maximum speed is doubled while the amplitude remains constant.

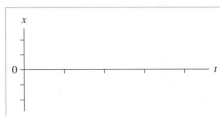

14.5 Pendulum Motion

14. The graph shows the displacement s versus time for an oscillating pendulum.

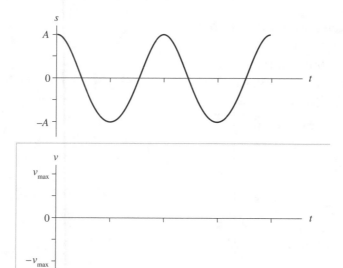

a. Draw the pendulum's velocity-versus-time graph.

b. In the space below, draw a *picture* of the pendulum that shows (and labels!)
   • The extremes of its motion.
   • Its position at $t = 0$ s.

14.6 Damped Oscillations

15. If the time constant τ of an oscillator is decreased,

    a. Is the drag force increased or decreased?

    b. Do the oscillations damp out more quickly or less quickly?

    c. Is τ greater or less than the time it takes for the oscillation amplitude to decrease to half of its original value?

    d. After the amplitude has decayed to half of its original value, how much longer will it take for the amplitude to decay to one-fourth its original value?

16. The figure below shows the envelope of the oscillations of a damped oscillator. On the same axes, draw the envelope of oscillations if

    a. The time constant is doubled.

    b. The time constant is halved.

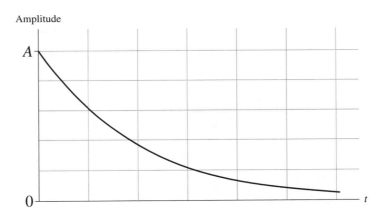

17. A block on a spring oscillates horizontally on a table *with friction*. Draw and label force vectors on the block to show all *horizontal* forces on the block.

    a. The mass is to the right of the equilibrium point and moving away from it.

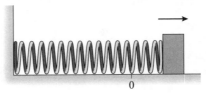

    b . The mass is to the right of the equilibrium point and approaching it.

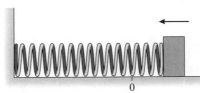

14.7 Driven Oscillations and Resonance

18. A car drives along a bumpy road on which the bumps are equally spaced. At a speed of 20 mph, the frequency of hitting bumps is equal to the natural frequency of the car bouncing on its springs.

    a. Draw a graph of the car's vertical bouncing amplitude as a function of its speed if the car has new shock absorbers (large damping coefficient).

    b. Draw a graph of the car's vertical bouncing amplitude as a function of its speed if the car has worn out shock absorbers (small damping coefficient).

    Draw both graphs on the same axes, and label them as to which is which.

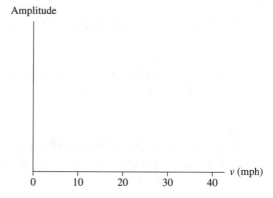

15 Traveling Waves and Sound

15.1 The Wave Model

15.2 Traveling Waves

1. The figure shows a wave pulse on a string at time $t = 0$ s. The pulse is traveling to the right at 1000 cm/s, or 1 cm/ms. There is a small bead attached to the string at $x = 9$ cm.

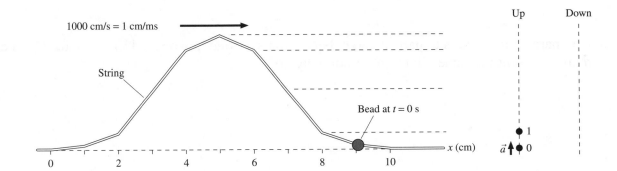

a. Draw a motion diagram of the bead. Show one frame every millisecond, from $t = 0$ ms to $t = 8$ ms. Show upward motion of the bead on the left dotted line and downward motion of the bead on the right line. Show the very top position, at $t = 4$ ms, on *both* lines. The first two frames, at 0 and 1 ms, are already shown and labeled.

b. Draw and label the $\vec{v}$ and $\vec{a}$ vectors on your motion diagram. Write $\vec{a} = \vec{0}$ if appropriate. Recall that velocity vectors go *between* the dots and acceleration vectors, which relate two velocity vectors, go *beside* the dots. The first acceleration vector is already shown.

c. At each instant, the bead feels two tension forces $\vec{T}_R$ and $\vec{T}_L$ along the direction of the string. Draw nine free-body diagrams, from $t = 0$ ms to $t = 8$ ms, showing the two tension forces *and* the net force $\vec{F}_{net}$. Use a different color for $\vec{F}_{net}$. If $\vec{F}_{net} = 0$, write that beside the figure. The $t = 1$ ms drawing is given as an example.

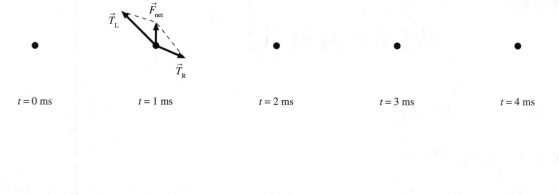

$t = 0$ ms $t = 1$ ms $t = 2$ ms $t = 3$ ms $t = 4$ ms

$t = 5$ ms $t = 6$ ms $t = 7$ ms $t = 8$ ms

d. Compare your answers to parts b and c. Is Newton's second law obeyed? That is, do $\vec{F}_{net}$ and $\vec{a}$ always point the same direction? If not, why not?

15.3 Graphical and Mathematical Descriptions of Waves

2. Each figure below shows a snapshot graph at time $t = 0$ s of a wave pulse on a string. The pulse on the left is traveling to the right at 100 cm/s; the pulse on the right is traveling to the left at 100 cm/s. Draw snapshot graphs of the wave pulse at the times shown next to the axes.

a.

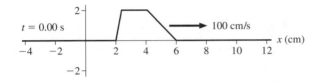

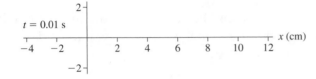

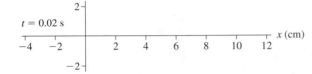

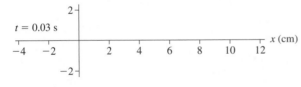

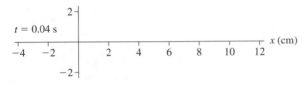

b.

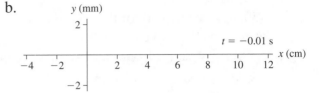

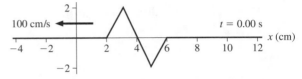

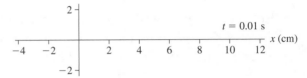

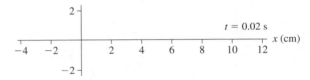

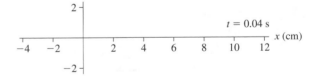

3. This snapshot graph is taken from Exercise 2a. On the axes below, draw the *history* graphs $y(x = 2$ cm$, t)$ and $y(x = 6$ cm$, t)$ showing the displacement at $x = 2$ cm and $x = 6$ cm as functions of time. Refer to your graphs in Exercise 2a to see what is happening at different instants of time.

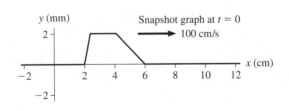

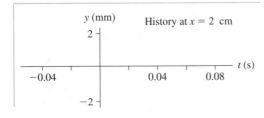

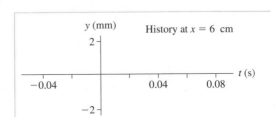

4. This snapshot graph is from Exercise 2b.

   a. Draw the history graph $y(x = 0 \text{ cm}, t)$ for this wave at the point $x = 0$ cm.

   b. Draw the *velocity*-versus-time graph for the piece of the string at $x = 0$ cm. Imagine painting a dot on the string at $x = 0$ cm. What is the velocity of this dot as a function of time as the wave passes by?

   c. As a wave passes through a medium, is the speed of a particle in the medium the same as or different from the speed of the wave through the medium? Explain.

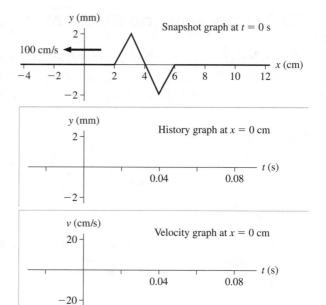

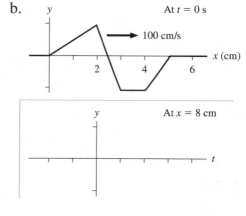

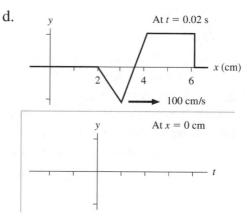

5. Below are four snapshot graphs of wave pulses on a string. For each, draw the history graph at the specified point on the x-axis. No time scale is provided on the t-axis, so you must determine an appropriate time scale and label the t-axis appropriately.

a.

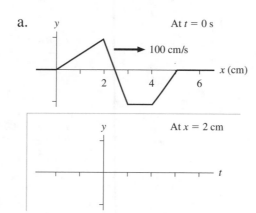

b.

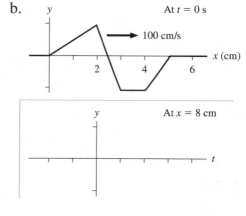

c.

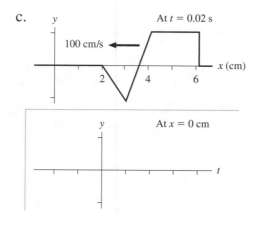

d.

6. A history graph $y(x = 0 \text{ cm}, t)$ is shown for the $x = 0$ cm point on a string. The pulse is moving to the right at 100 cm/s.

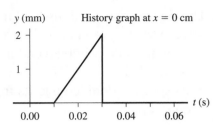

History graph at $x = 0$ cm

a. Does the $x = 0$ cm point on the string rise quickly and then fall slowly, or rise slowly and then fall quickly? Explain.

b. At what time does the leading edge of the wave pulse arrive at $x = 0$ cm? _____

c. At $t = 0$ s, how far is the leading edge of the wave pulse from $x = 0$ cm? Explain.

d. At $t = 0$ s, is the leading edge to the right or to the left of $x = 0$ cm? _____

e. At what time does the trailing edge of the wave pulse leave $x = 0$ cm? _____

f. At $t = 0$ s, how far is the trailing edge of the pulse from $x = 0$ cm? Explain.

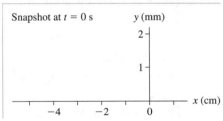

Snapshot at $t = 0$ s

g. By referring to the answers you've just given, draw a snapshot graph $y(x = 0 \text{ s})$ showing the wave pulse on the string at $t = 0$ s.

7. Below are three history graphs for wave pulses on a string. The speed and direction of each pulse are indicated. For each, draw the snapshot graph at the specified instant of time. No distance scale is provided, so you must determine an appropriate scale and label the x-axis appropriately.

a.
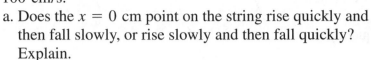
History at $x = 0$ cm
100 cm/s to the right

b.
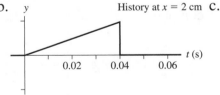
History at $x = 2$ cm
100 cm/s to the left

c.
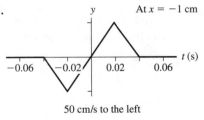
At $x = -1$ cm
50 cm/s to the left

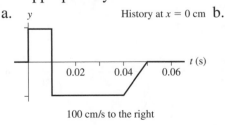
Snapshot at $t = 0.05$ s

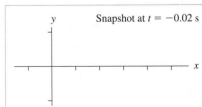
Snapshot at $t = -0.02$ s

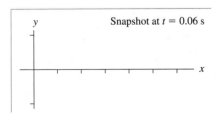
Snapshot at $t = 0.06$ s

8. The figure shows a sinusoidal traveling wave. Draw a graph of the wave if:
   a. Its amplitude is halved and its wavelength is doubled.
   b. Its speed is doubled and its frequency is quadrupled.

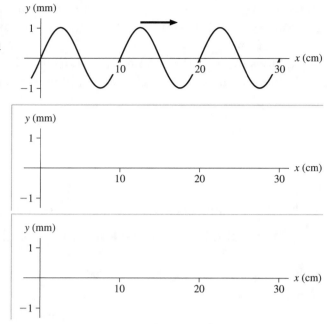

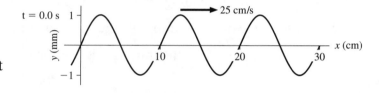

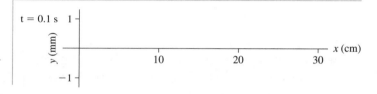

9. The wave shown at time $t = 0$ s is traveling to the right at a speed of 25 cm/s.
   a. Draw snapshot graphs of this wave at times $t = 0.1$ s, $t = 0.2$ s, $t = 0.3$ s, and $t = 0.4$ s. (Draw on axes provided.)
   b. What is the wavelength of the wave? _____
   c. Based on your graphs, what is the period of the wave? _____
   d. What is the frequency of the wave?

   _____

   e. What is the value of the product λf?

   _____

   f. How does this value of λf compare to the speed of the wave?

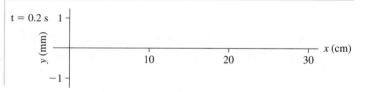

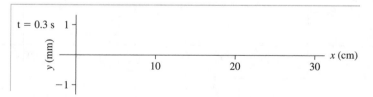

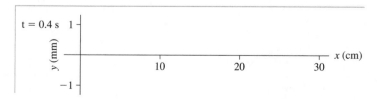

15.4 Sound and Light Waves

10. A horizontal Slinky is at rest on a table. A wave pulse is sent along the Slinky, causing the top of link 5 to move *horizontally* with the displacement shown in the graph.

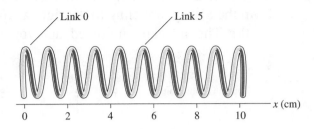

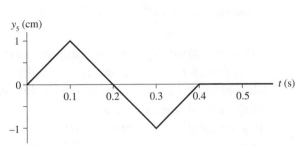

   a. Is this a transverse or a longitudinal wave? Explain.

   b. What is the position of link 5

      at $t = 0.1\,s$? _____

      What is the position of link 5

      at $t = 0.2\,s$? _____

      What is the position of link 5

      at $t = 0.3\,s$? _____

   Note: *Position*, not displacement.

   c. Draw a velocity-versus-time graph of link 5. Add an appropriate scale to the vertical axis.

   d. Can you determine, from the information given, whether the wave pulse is traveling to the right or to the left? If so, give the direction and explain how you found it. If not, why not?

   e. Can you determine, from the information given, the speed of the wave? If so, give the speed and explain how you found it. If not, why not?

11. We can use a series of dots to represent the positions of the links in a Slinky. The top set of dots shows a Slinky in equilibrium with a 1 cm spacing between the links. A wave pulse is sent down the Slinky, traveling to the right at 10 cm/s. The second set of dots shows the Slinky at $t = 0$ s. The links are numbered, and you can measure the displacement Δx of each link.

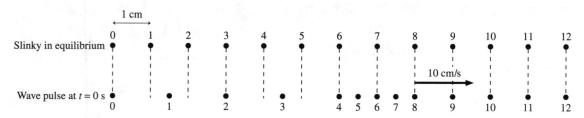

a. Draw a snapshot graph showing the displacement of each link at $t = 0$ s. There are 13 links, so your graph should have 13 dots. Connect your dots with lines to make a continuous graph.

b. Is your graph a "picture" of the wave or a "representation" of the wave? Explain.

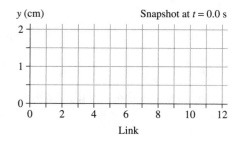

c. Which links are in compression? (list their numbers) _____

Which links are in rarefaction? (list their numbers) _____

d. Draw graphs of displacement versus the link number at $t = 0.1$ s and $t = 0.2$ s.

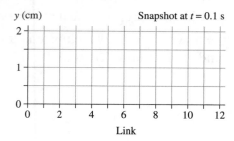 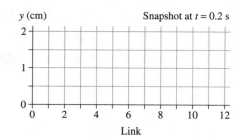

e. Now draw dot pictures of the links at $t = 0.1$ s and $t = 0.2$ s. The equilibrium positions and the $t = 0$ s picture are shown for reference.

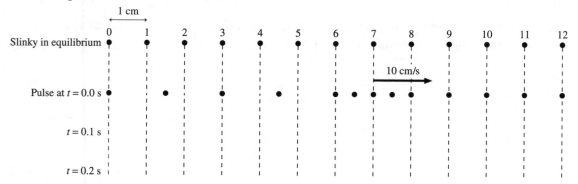

15.5 Energy and Intensity

15.6 Loudness of Sound

12. A wave-front diagram is shown for a sinusoidal plane wave at time $t = 0$ s. The diagram shows only the xy-plane, but the wave extends above and below the plane of the paper.

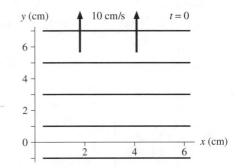

   a. What is the wavelength of this wave? _____

   b. At $t = 0$ s, for which values of y is the wave a crest?

   c. At $t = 0$ s, for which values of y is the wave a trough?

   d. Can you tell if this is a transverse or a longitudinal wave? If so, which is it and how did you determine it? If not, why not?

   e. How does the displacement at the point $(x, y, z) = (6, 5, 0)$ compare to the displacement at the point $(2, 5, 0)$? Is it more, less, the same, or is there no way to tell? Explain.

   f. How does the displacement at the point $(x, y, z) = (2, 5, 5)$ compare to the displacement at the point $(2, 5, 0)$? Is it more, less, the same, or is there no way to tell? Explain.

   g. On the left axes below, draw a snapshot graph of the displacement $D(y, t = 0$ s$)$ along the y-axis at $t = 0$ s.

   h. On the right axes below, draw a wave-front diagram at time $t = 0.3$ s.

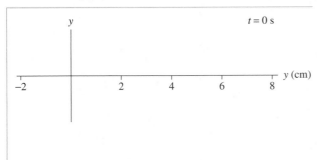

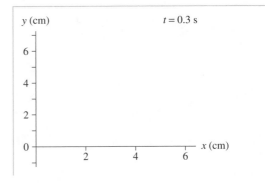

13. The figure shows the path of a light wave past a series of equally spaced *transparent* grids. The portion of the first two grids that would be illuminated by the light is represented by the shaded area in the first two grids.

   a. Complete the figure by shading in the spaces that would be illuminated in the remaining two grids.

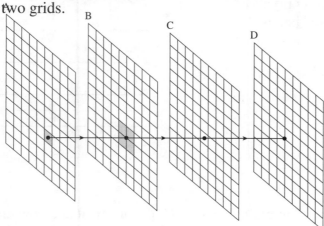

   b. How does the energy of the light wave passing through the first transparent grid (A) compare to the energy of the light wave passing through the fourth transparent grid (D)? Explain.

   c. How does the intensity of the light in the fourth grid (D) compare to the intensity of the light through the first grid (A)? Explain.

15.7 The Doppler Effect and Shock Waves

14. Five expanding wave fronts from a moving sound source
are shown. The dots represent the centers of the respective
circular wave fronts, which is the location of the source
when that wave front was emitted. The frequency of the
sound emitted by the source is constant.

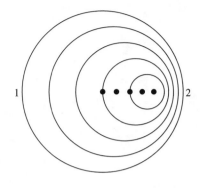

a. Indicate on the figure the direction of motion of the
source. Which sound wave front was produced first?
How do you know? Explain.

b. Do the observers at locations 1 and 2 hear the same frequency of sound? If not, which one
experiences a higher frequency? Explain.

c. Assume that the sound wave you identified in part a marks the beginning of the sound. Do
the observers at 1 and 2 first hear the sound at the same time? If not, which one hears the
sound first? Explain.

d. If the speed of sound in this medium is v, what is the speed of the source? Explain.

15. You are standing at $x = 0$ m, listening to a sound that is
emitted at frequency f_0. At $t = 0$ s, the sound source is at
$x = 20$ m and moving toward you at a steady 10 m/s.
Draw a graph showing the frequency you hear from
$t = 0$ s to $t = 4$ s. Only the shape of the graph is
important, not the numerical values of f.

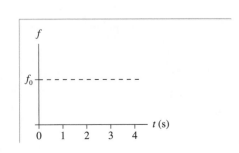

16 Superposition and Standing Waves

16.1 The Principle of Superposition

1. Two pulses on a string are approaching each other at 10 m/s. Draw snapshot graphs of the string at the three times indicated.

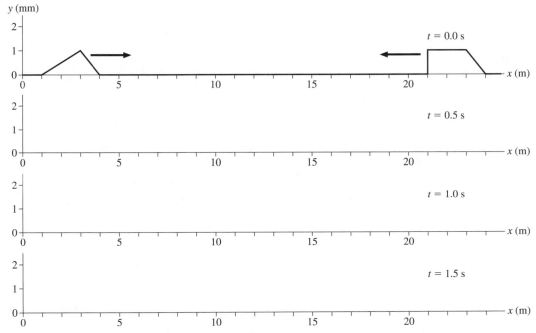

2. Two pulses on a string are approaching each other at 10 m/s. Draw a snapshot graph of the string at $t = 1$ s.

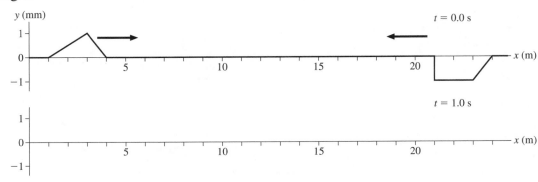

16.2 Standing Waves

3. Two waves are traveling in opposite directions along a string. Each has a speed of 1000 cm/s, or 1 cm/ms, and an amplitude of 1 cm. The first set of graphs below shows each wave at $t = 0$ ms.

   a. On the axes at the right, draw the superposition of these two waves at $t = 0$ ms.

   b. On the axes at the left, draw each of the two displacements every 1 ms until $t = 8$ ms.
      The waves extend beyond the graph edges, so new pieces of the wave will move in.

   c. On the axes at the right, draw the superposition of the two waves at the same instant.

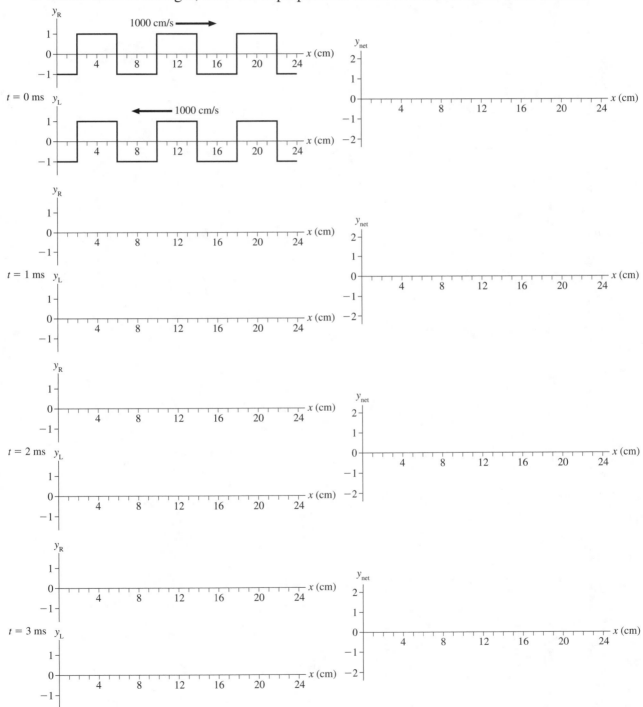

(Continues next page)

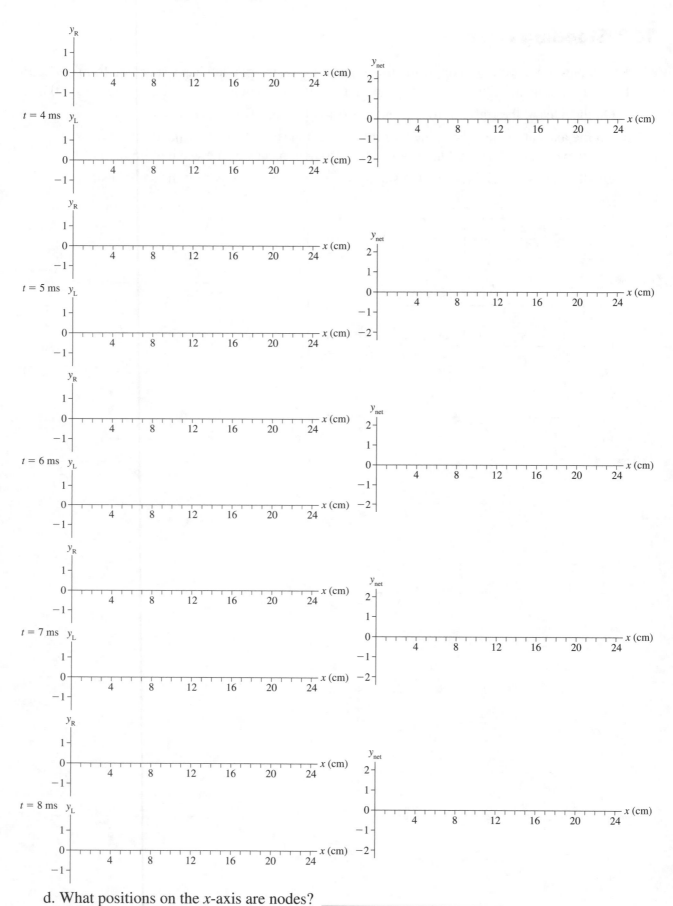

d. What positions on the *x*-axis are nodes? _____

e. What positions on the *x*-axis are antinodes? _____

16.3 Standing Waves on Strings

4. This standing wave has a period of 8 ms. Draw snapshot graphs of the string every 1 ms from $t = 1$ ms to $t = 8$ ms. Think carefully about the proper amplitude at each instant.

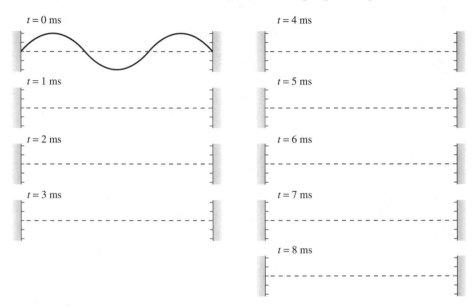

5. The figure shows a standing wave on a string. It has frequency f.

a. Draw the standing wave if the frequency is changed to $\frac{2}{3}f$ and to $\frac{3}{2}f$.

Original wave, frequency f Frequency $\frac{2}{3}f$ Frequency $\frac{3}{2}f$

b. Is there a standing wave if the frequency is changed to $\frac{1}{4}f$? If so, how many antinodes does it have? If not, why not?

6. The figure shows a standing wave on a string.

a. Draw the standing wave if the tension is quadrupled while the frequency is held constant.

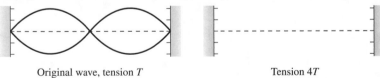

Original wave, tension T Tension $4T$

b. Suppose the tension is merely doubled while the frequency remains constant. Will there be a standing wave? If so, how many antinodes will it have? If not, why not?

16.4 Standing Sound Waves

7. The picture shows a standing sound wave in a 32.-mm-long tube of air that is open at both ends in the $m = 2$ mode. To help you visualize the horizontal motion of the air molecules in this standing wave, on the next page are nine graphs, every one-eighth of a period from $t = 0$ to $t = T$. Each graph represents the displacements at that instant of the time of the molecules in the 32-mm-long tube. Positive values are displacements to the right, negative values are displacements to the left.

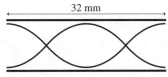

a. Consider nine air molecules that, in equilibrium, are 4 mm apart and lie along the axis of the tube. The top picture on the right shows these molecules in their equilibrium positions. The dotted lines down the page—spaced 4 mm apart—are reference lines showing the equilibrium positions. Read each graph carefully, then draw nine dots to show the positions of the nine air molecules at each instant of time. The first one, for $t = 0$, has already been done to illustrate the procedure.

   Note: It's a good approximation to assume that the left dot moves in the pattern 4, 3, 0, −3, −4, −3, 0, 3, 4 mm; the second dot in the pattern 3, 2, 0, −2, −3, −2, 0, 2, 3 mm; and so on.

b. At what times does the air reach maximum compression, and where does it occur?

   Max compression at time _____   Max compression at position _____

   _____   _____

   _____   _____

c. What is the relationship between the positions of maximum compression and the nodes of the standing wave?

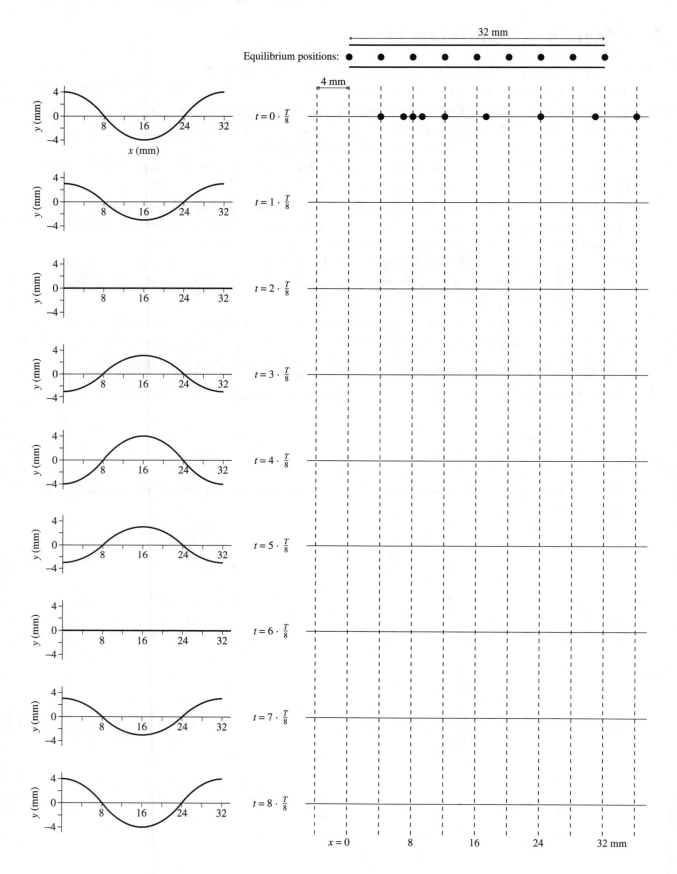

16.5 Speech and Hearing

16.6 The Interference of Waves from Two Sources

8. The figure shows a snapshot graph at $t = 0$ s of loudspeakers emitting triangular-shaped sound waves. Speaker 2 can be moved forward or backward along the axis. Both speakers vibrate in phase at the same frequency. The second speaker is drawn below the first, so that the figure is clear, but you want to think of the two waves as overlapped as they travel along the x-axis.

   a. On the left set of axes, draw the $t = 0$ s snapshot graph of the second wave if speaker 2 is placed at each of the positions shown. The first graph, with $x_{speaker} = 2$ m, is already drawn.

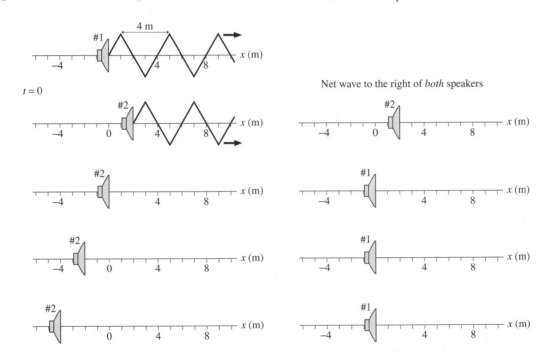

   b. On the right set of axes, draw the superposition $\Delta p_{net} = \Delta p_1 + \Delta p_2$ of the waves from the two speakers. Δp_{net} exists only to the right of *both* speakers. It is the net wave traveling to the right.

   c. What separations between the speakers give constructive interference? _____

   d. What are the $\Delta x/\lambda$ ratios at the points of constructive interference? _____

   e. What separations between the speakers give destructive interference? _____

   f. What are the $\Delta x/\lambda$ ratios at the points of destructive interference? _____

9. Consider the two loudspeakers of Exercise 8.

   a. Copy the speaker 1 and 2 graphs from Exercise 8 onto the first set of axes below for the situation in which speaker 2 is 4 m behind speaker 1. Then draw their superposition on the axes at the right. This simply repeats your last set of graphs from Exercise 8.

   b. On the axes on the left, draw snapshot graphs of the two waves at times $t = \frac{1}{4}T, \frac{2}{4}T$, and $\frac{3}{4}T$, where T is the wave's period.

   c. On the right axes, draw the superposition of the two waves.

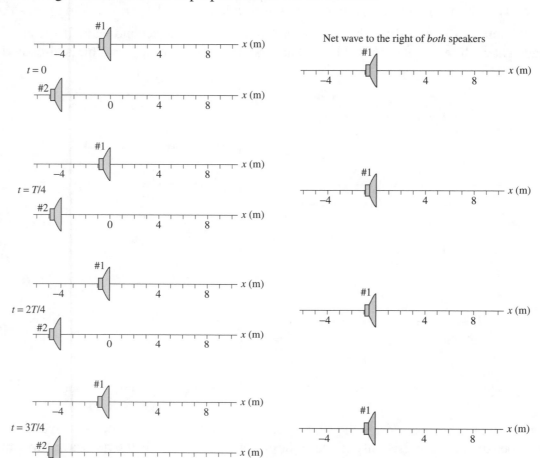

   d. Is the net wave a traveling wave or a standing wave? Use your *observations* to explain.

10. Speakers 1 and 2 are 12 m apart. Both emit identical triangular sound waves with $\lambda = 4$ m and are in phase initially at their respective locations. Point A is $r_1 = 16$ m from speaker 1.

    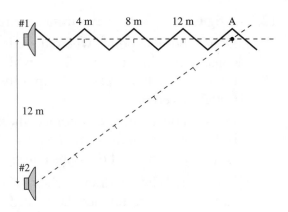

    a. What is distance r_2 from speaker 2 to A?

    _____

    b. Draw the wave from speaker 2 along the dashed line to just past point A.

    c. At A, is wave 1 a crest, trough, or zero? _____

       At A, is wave 2 a crest, trough, or zero? _____

    d. What is the path length difference $\Delta r = r_2 - r_1$? _____ What is the ratio $\Delta r/\lambda$? _____

    e. Is the interference at point A constructive, destructive, or in between? _____

11. Speakers 1 and 2 are 18 m apart. Both emit identical triangular sound waves with $l = 4$ m and are in phase initially at their respective locations. Point B is r1 = 24 m from speaker 1.
    a. What is distance r_2 from speaker 2 to B?

    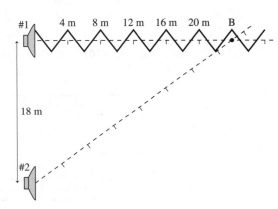

    _____

    b. Draw the wave from speaker 2 along the dashed line to just past point A.

    c. At B, is wave 1 a crest, trough, or zero? _____

       At B, is wave 2 a crest, trough, or zero? _____

    d. What is the path length difference $\Delta r = r_2 - r_1$? _____ What is the ratio $\Delta r/\lambda$? _____

    e. Is the interference at point B constructive, destructive, or in between? _____

12. The figure shows the wave-front pattern emitted by two loudspeakers.

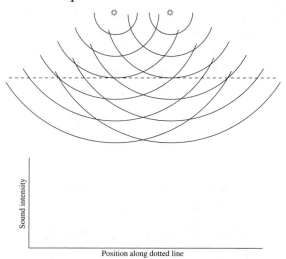

   a. Draw a dot • at points where there is constructive interference. These will be points where two crests overlap *or* two troughs overlap.

   b. Draw an open circle ○ at points where there is destructive interference. These will be points where a crest overlaps a trough.

   c. Use a **black** line to draw each "ray" of constructive interference. Use a **red** line to draw each "ray" of destructive interference.

   d. Draw a graph on the axes above of the sound intensity you would hear if you walked along the horizontal dotted line. Use the same horizontal scale as the figure so that your graph lines up with the figure above it.

16.7 Beats

13. The two waves arrive simultaneously at a point in space from two different sources.

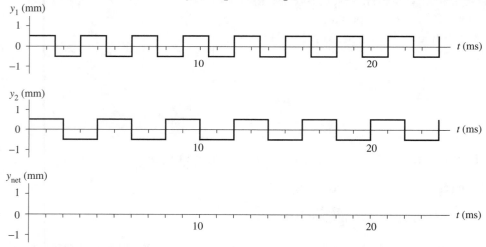

   a. Period of wave 1? _____   Frequency of wave 1? _____

   b. Period of wave 2? _____   Frequency of wave 2? _____

   c. Draw the graph of the net wave at this point on the third set of axes. Be accurate, use a ruler!

   d. Period of the net wave? _____   Frequency of the net wave? _____

   e. Is the frequency of the superposition what you would expect as a beat frequency? Explain.

DYNAMICS WORKSHEET   Name _____   Problem _____

PREPARE
- List knowns. Identify what you're trying to find.
- Identify forces and draw a free-body diagram.

- Draw a pictorial representation for problems with motion: Show important points in the motion, establish a coordinate system, define symbols, draw a motion diagram.

Known

Find

SOLVE
Start with Newton's first or second law in component form, adding other information as needed to solve the problem.

ASSESS
Have you answered the question?
Do you have correct units, signs, and significant figures?
Is your answer reasonable?

DYNAMICS WORKSHEET   Name _____   Problem _____

PREPARE

- List knowns. Identify what you're trying to find.
- Identify forces and draw a free-body diagram.

- Draw a pictorial representation for problems with motion: Show important points in the motion, establish a coordinate system, define symbols, draw a motion diagram.

Known

Find

SOLVE

Start with Newton's first or second law in component form, adding other information as needed to solve the problem.

ASSESS

Have you answered the question?

Do you have correct units, signs, and significant figures?

Is your answer reasonable?

DYNAMICS WORKSHEET Name _____ Problem _____

PREPARE

- List knowns. Identify what you're trying to find.
- Identify forces and draw a free-body diagram.

- Draw a pictorial representation for problems with motion: Show important points in the motion, establish a coordinate system, define symbols, draw a motion diagram.

Known

Find

SOLVE

Start with Newton's first or second law in component form, adding other information as needed to solve the problem.

ASSESS

Have you answered the question?

Do you have correct units, signs, and significant figures?

Is your answer reasonable?

DYNAMICS WORKSHEET   Name _____   Problem _____

PREPARE

- List knowns. Identify what you're trying to find.
- Identify forces and draw a free-body diagram.

- Draw a pictorial representation for problems with motion: Show important points in the motion, establish a coordinate system, define symbols, draw a motion diagram.

Known

Find

SOLVE

Start with Newton's first or second law in component form, adding other information as needed to solve the problem.

ASSESS

Have you answered the question?

Do you have correct units, signs, and significant figures?

Is your answer reasonable?

DYNAMICS WORKSHEET Name _____ Problem _____

PREPARE

- List knowns. Identify what you're trying to find.
- Identify forces and draw a free-body diagram.

- Draw a pictorial representation for problems with motion: Show important points in the motion, establish a coordinate system, define symbols, draw a motion diagram.

Known

Find

SOLVE

Start with Newton's first or second law in component form, adding other information as needed to solve the problem.

ASSESS

Have you answered the question?

Do you have correct units, signs, and significant figures?

Is your answer reasonable?

DYNAMICS WORKSHEET Name _____ Problem _____

PREPARE

- List knowns. Identify what you're trying to find.
- Identify forces and draw a free-body diagram.

- Draw a pictorial representation for problems with motion: Show important points in the motion, establish a coordinate system, define symbols, draw a motion diagram.

Known

Find

SOLVE

Start with Newton's first or second law in component form, adding other information as needed to solve the problem.

ASSESS

Have you answered the question?
Do you have correct units, signs, and significant figures?
Is your answer reasonable?

DYNAMICS WORKSHEET Name _____ Problem _____

PREPARE
- List knowns. Identify what you're trying to find.
- Identify forces and draw a free-body diagram.

- Draw a pictorial representation for problems with motion: Show important points in the motion, establish a coordinate system, define symbols, draw a motion diagram.

Known

Find

SOLVE
Start with Newton's first or second law in component form, adding other information as needed to solve the problem.

ASSESS
Have you answered the question?
Do you have correct units, signs, and significant figures?
Is your answer reasonable?

DYNAMICS WORKSHEET   Name _____   Problem _____

PREPARE

- List knowns. Identify what you're trying to find.
- Identify forces and draw a free-body diagram.

- Draw a pictorial representation for problems with motion: Show important points in the motion, establish a coordinate system, define symbols, draw a motion diagram.

Known

Find

SOLVE

Start with Newton's first or second law in component form, adding other information as needed to solve the problem.

ASSESS

Have you answered the question?
Do you have correct units, signs, and significant figures?
Is your answer reasonable?

DYNAMICS WORKSHEET   Name _____   Problem _____

PREPARE

- List knowns. Identify what you're trying to find.
- Identify forces and draw a free-body diagram.

- Draw a pictorial representation for problems with motion: Show important points in the motion, establish a coordinate system, define symbols, draw a motion diagram.

Known

Find

SOLVE

Start with Newton's first or second law in component form, adding other information as needed to solve the problem.

ASSESS

Have you answered the question?

Do you have correct units, signs, and significant figures?

Is your answer reasonable?

DYNAMICS WORKSHEET   Name _____   Problem _____

PREPARE

- List knowns. Identify what you're trying to find.
- Identify forces and draw a free-body diagram.

- Draw a pictorial representation for problems with motion: Show important points in the motion, establish a coordinate system, define symbols, draw a motion diagram.

Known

Find

SOLVE

Start with Newton's first or second law in component form, adding other information as needed to solve the problem.

ASSESS

Have you answered the question?

Do you have correct units, signs, and significant figures?

Is your answer reasonable?

DYNAMICS WORKSHEET Name _____ Problem _____

PREPARE
- List knowns. Identify what you're trying to find.
- Identify forces and draw a free-body diagram.

- Draw a pictorial representation for problems with motion: Show important points in the motion, establish a coordinate system, define symbols, draw a motion diagram.

Known

Find

SOLVE
Start with Newton's first or second law in component form, adding other information as needed to solve the problem.

ASSESS
Have you answered the question?
Do you have correct units, signs, and significant figures?
Is your answer reasonable?

DYNAMICS WORKSHEET   Name _____   Problem _____

PREPARE
- List knowns. Identify what you're trying to find.
- Identify forces and draw a free-body diagram.

- Draw a pictorial representation for problems with motion: Show important points in the motion, establish a coordinate system, define symbols, draw a motion diagram.

Known

Find

SOLVE
Start with Newton's first or second law in component form, adding other information as needed to solve the problem.

ASSESS
Have you answered the question?
Do you have correct units, signs, and significant figures?
Is your answer reasonable?

DYNAMICS WORKSHEET Name _____ Problem _____

- List knowns. Identify what you're trying to find.
- Identify forces and draw a free-body diagram.

- Draw a pictorial representation for problems with motion: Show important points in the motion, establish a coordinate system, define symbols, draw a motion diagram.

Known

Find

SOLVE
Start with Newton's first or second law in component form, adding other information as needed to solve the problem.

ASSESS
Have you answered the question?
Do you have correct units, signs, and significant figures?
Is your answer reasonable?

DYNAMICS WORKSHEET Name _____ Problem _____

PREPARE

- List knowns. Identify what you're trying to find.
- Identify forces and draw a free-body diagram.

- Draw a pictorial representation for problems with motion: Show important points in the motion, establish a coordinate system, define symbols, draw a motion diagram.

Known

Find

SOLVE

Start with Newton's first or second law in component form, adding other information as needed to solve the problem.

ASSESS

Have you answered the question?
Do you have correct units, signs, and significant figures?
Is your answer reasonable?

DYNAMICS WORKSHEET   Name _____   Problem _____

PREPARE

- List knowns. Identify what you're trying to find.
- Identify forces and draw a free-body diagram.

- Draw a pictorial representation for problems with motion: Show important points in the motion, establish a coordinate system, define symbols, draw a motion diagram.

Known

Find

SOLVE

Start with Newton's first or second law in component form, adding other information as needed to solve the problem.

ASSESS

Have you answered the question?

Do you have correct units, signs, and significant figures?

Is your answer reasonable?

DYNAMICS WORKSHEET Name _____ Problem _____

PREPARE

- List knowns. Identify what you're trying to find.
- Identify forces and draw a free-body diagram.

- Draw a pictorial representation for problems with motion: Show important points in the motion, establish a coordinate system, define symbols, draw a motion diagram.

Known

Find

SOLVE

Start with Newton's first or second law in component form, adding other information as needed to solve the problem.

ASSESS

Have you answered the question?

Do you have correct units, signs, and significant figures?

Is your answer reasonable?

DYNAMICS WORKSHEET   Name _____   Problem _____

PREPARE

- List knowns. Identify what you're trying to find.
- Identify forces and draw a free-body diagram.

- Draw a pictorial representation for problems with motion: Show important points in the motion, establish a coordinate system, define symbols, draw a motion diagram.

Known

Find

SOLVE

Start with Newton's first or second law in component form, adding other information as needed to solve the problem.

ASSESS

Have you answered the question?

Do you have correct units, signs, and significant figures?

Is your answer reasonable?

DYNAMICS WORKSHEET   Name _____ Problem _____

PREPARE

- List knowns. Identify what you're trying to find.
- Identify forces and draw a free-body diagram.

- Draw a pictorial representation for problems with motion: Show important points in the motion, establish a coordinate system, define symbols, draw a motion diagram.

Known

Find

SOLVE

Start with Newton's first or second law in component form, adding other information as needed to solve the problem.

ASSESS

Have you answered the question?

Do you have correct units, signs, and significant figures?

Is your answer reasonable?

DYNAMICS WORKSHEET Name _____ Problem _____

PREPARE

- List knowns. Identify what you're trying to find.
- Identify forces and draw a free-body diagram.

- Draw a pictorial representation for problems with motion: Show important points in the motion, establish a coordinate system, define symbols, draw a motion diagram.

Known

Find

SOLVE

Start with Newton's first or second law in component form, adding other information as needed to solve the problem.

ASSESS

Have you answered the question?
Do you have correct units, signs, and significant figures?
Is your answer reasonable?

DYNAMICS WORKSHEET   Name _____   Problem _____

PREPARE

- List knowns. Identify what you're trying to find.
- Identify forces and draw a free-body diagram.

- Draw a pictorial representation for problems with motion: Show important points in the motion, establish a coordinate system, define symbols, draw a motion diagram.

Known

Find

SOLVE

Start with Newton's first or second law in component form, adding other information as needed to solve the problem.

ASSESS

Have you answered the question?

Do you have correct units, signs, and significant figures?

Is your answer reasonable?

DYNAMICS WORKSHEET Name _____ Problem _____

PREPARE

- List knowns. Identify what you're trying to find.
- Identify forces and draw a free-body diagram.

- Draw a pictorial representation for problems with motion: Show important points in the motion, establish a coordinate system, define symbols, draw a motion diagram.

Known

Find

SOLVE

Start with Newton's first or second law in component form, adding other information as needed to solve the problem.

ASSESS

Have you answered the question?

Do you have correct units, signs, and significant figures?

Is your answer reasonable?

DYNAMICS WORKSHEET   Name _____   Problem _____

PREPARE

- List knowns. Identify what you're trying to find.
- Identify forces and draw a free-body diagram.

- Draw a pictorial representation for problems with motion: Show important points in the motion, establish a coordinate system, define symbols, draw a motion diagram.

Known

Find

SOLVE

Start with Newton's first or second law in component form, adding other information as needed to solve the problem.

ASSESS

Have you answered the question?
Do you have correct units, signs, and significant figures?
Is your answer reasonable?

DYNAMICS WORKSHEET Name _____ Problem _____

PREPARE

- List knowns. Identify what you're trying to find.
- Identify forces and draw a free-body diagram.

- Draw a pictorial representation for problems with motion: Show important points in the motion, establish a coordinate system, define symbols, draw a motion diagram.

Known

Find

SOLVE

Start with Newton's first or second law in component form, adding other information as needed to solve the problem.

ASSESS

Have you answered the question?
Do you have correct units, signs, and significant figures?
Is your answer reasonable?

DYNAMICS WORKSHEET Name _____ Problem _____

PREPARE

- List knowns. Identify what you're trying to find.
- Identify forces and draw a free-body diagram.

- Draw a pictorial representation for problems with motion: Show important points in the motion, establish a coordinate system, define symbols, draw a motion diagram.

Known

Find

SOLVE

Start with Newton's first or second law in component form, adding other information as needed to solve the problem.

ASSESS

Have you answered the question?
Do you have correct units, signs, and significant figures?
Is your answer reasonable?

DYNAMICS WORKSHEET Name _____ Problem _____

PREPARE

- List knowns. Identify what you're trying to find.
- Identify forces and draw a free-body diagram.

- Draw a pictorial representation for problems with motion: Show important points in the motion, establish a coordinate system, define symbols, draw a motion diagram.

Known

Find

SOLVE

Start with Newton's first or second law in component form, adding other information as needed to solve the problem.

ASSESS

Have you answered the question?

Do you have correct units, signs, and significant figures?

Is your answer reasonable?

DYNAMICS WORKSHEET   Name _____   Problem _____

PREPARE

- List knowns. Identify what you're trying to find.
- Identify forces and draw a free-body diagram.

- Draw a pictorial representation for problems with motion: Show important points in the motion, establish a coordinate system, define symbols, draw a motion diagram.

Known

Find

SOLVE

Start with Newton's first or second law in component form, adding other information as needed to solve the problem.

ASSESS

Have you answered the question?

Do you have correct units, signs, and significant figures?

Is your answer reasonable?

DYNAMICS WORKSHEET Name _____ Problem _____

PREPARE

- List knowns. Identify what you're trying to find.
- Identify forces and draw a free-body diagram.

- Draw a pictorial representation for problems with motion: Show important points in the motion, establish a coordinate system, define symbols, draw a motion diagram.

Known

Find

SOLVE

Start with Newton's first or second law in component form, adding other information as needed to solve the problem.

ASSESS

Have you answered the question?

Do you have correct units, signs, and significant figures?

Is your answer reasonable?

DYNAMICS WORKSHEET   Name _____   Problem _____

PREPARE

- List knowns. Identify what you're trying to find.
- Identify forces and draw a free-body diagram.

- Draw a pictorial representation for problems with motion: Show important points in the motion, establish a coordinate system, define symbols, draw a motion diagram.

Known

Find

SOLVE

Start with Newton's first or second law in component form, adding other information as needed to solve the problem.

ASSESS

Have you answered the question?

Do you have correct units, signs, and significant figures?

Is your answer reasonable?

DYNAMICS WORKSHEET   Name _____   Problem _____

PREPARE

- List knowns. Identify what you're trying to find.
- Identify forces and draw a free-body diagram.

- Draw a pictorial representation for problems with motion: Show important points in the motion, establish a coordinate system, define symbols, draw a motion diagram.

Known

Find

SOLVE

Start with Newton's first or second law in component form, adding other information as needed to solve the problem.

ASSESS

Have you answered the question?

Do you have correct units, signs, and significant figures?

Is your answer reasonable?

DYNAMICS WORKSHEET   Name _____   Problem _____

PREPARE

- List knowns. Identify what you're trying to find.
- Identify forces and draw a free-body diagram.

- Draw a pictorial representation for problems with motion: Show important points in the motion, establish a coordinate system, define symbols, draw a motion diagram.

Known

Find

SOLVE

Start with Newton's first or second law in component form, adding other information as needed to solve the problem.

ASSESS

Have you answered the question?

Do you have correct units, signs, and significant figures?

Is your answer reasonable?

DYNAMICS WORKSHEET Name _____ Problem _____

PREPARE

- List knowns. Identify what you're trying to find.
- Identify forces and draw a free-body diagram.

- Draw a pictorial representation for problems with motion: Show important points in the motion, establish a coordinate system, define symbols, draw a motion diagram.

Known

Find

SOLVE

Start with Newton's first or second law in component form, adding other information as needed to solve the problem.

ASSESS

Have you answered the question?

Do you have correct units, signs, and significant figures?

Is your answer reasonable?

DYNAMICS WORKSHEET Name _____ Problem _____

PREPARE

- List knowns. Identify what you're trying to find.
- Identify forces and draw a free-body diagram.

- Draw a pictorial representation for problems with motion: Show important points in the motion, establish a coordinate system, define symbols, draw a motion diagram.

Known

Find

SOLVE

Start with Newton's first or second law in component form, adding other information as needed to solve the problem.

ASSESS

Have you answered the question?

Do you have correct units, signs, and significant figures?

Is your answer reasonable?

DYNAMICS WORKSHEET   Name _____   Problem _____

Known

Find

DYNAMICS WORKSHEET Name _____ Problem _____

PREPARE

- List knowns. Identify what you're trying to find.
- Identify forces and draw a free-body diagram.

- Draw a pictorial representation for problems with motion: Show important points in the motion, establish a coordinate system, define symbols, draw a motion diagram.

Known

Find

SOLVE

Start with Newton's first or second law in component form, adding other information as needed to solve the problem.

ASSESS

Have you answered the question?

Do you have correct units, signs, and significant figures?

Is your answer reasonable?

DYNAMICS WORKSHEET   Name _____   Problem _____

PREPARE

- List knowns. Identify what you're trying to find.
- Identify forces and draw a free-body diagram.

- Draw a pictorial representation for problems with motion: Show important points in the motion, establish a coordinate system, define symbols, draw a motion diagram.

Known

Find

SOLVE

Start with Newton's first or second law in component form, adding other information as needed to solve the problem.

ASSESS

Have you answered the question?

Do you have correct units, signs, and significant figures?

Is your answer reasonable?

DYNAMICS WORKSHEET   Name _____   Problem _____

PREPARE

- List knowns. Identify what you're trying to find.
- Identify forces and draw a free-body diagram.

- Draw a pictorial representation for problems with motion: Show important points in the motion, establish a coordinate system, define symbols, draw a motion diagram.

Known

Find

SOLVE

Start with Newton's first or second law in component form, adding other information as needed to solve the problem.

ASSESS

Have you answered the question?

Do you have correct units, signs, and significant figures?

Is your answer reasonable?

DYNAMICS WORKSHEET Name _____ Problem _____

PREPARE
- List knowns. Identify what you're trying to find.
- Identify forces and draw a free-body diagram.

- Draw a pictorial representation for problems with motion: Show important points in the motion, establish a coordinate system, define symbols, draw a motion diagram.

Known

Find

SOLVE
Start with Newton's first or second law in component form, adding other information as needed to solve the problem.

ASSESS
Have you answered the question?
Do you have correct units, signs, and significant figures?
Is your answer reasonable?

DYNAMICS WORKSHEET   Name _____   Problem _____

PREPARE

- List knowns. Identify what you're trying to find.
- Identify forces and draw a free-body diagram.

- Draw a pictorial representation for problems with motion: Show important points in the motion, establish a coordinate system, define symbols, draw a motion diagram.

Known

Find

SOLVE

Start with Newton's first or second law in component form, adding other information as needed to solve the problem.

ASSESS

Have you answered the question?
Do you have correct units, signs, and significant figures?
Is your answer reasonable?

DYNAMICS WORKSHEET Name _____ Problem _____

PREPARE

- List knowns. Identify what you're trying to find.
- Identify forces and draw a free-body diagram.

- Draw a pictorial representation for problems with motion: Show important points in the motion, establish a coordinate system, define symbols, draw a motion diagram.

Known

Find

SOLVE

Start with Newton's first or second law in component form, adding other information as needed to solve the problem.

ASSESS

Have you answered the question?

Do you have correct units, signs, and significant figures?

Is your answer reasonable?

DYNAMICS WORKSHEET   Name _____   Problem _____

PREPARE

- List knowns. Identify what you're trying to find.
- Identify forces and draw a free-body diagram.

- Draw a pictorial representation for problems with motion: Show important points in the motion, establish a coordinate system, define symbols, draw a motion diagram.

Known

Find

SOLVE

Start with Newton's first or second law in component form, adding other information as needed to solve the problem.

ASSESS

Have you answered the question?

Do you have correct units, signs, and significant figures?

Is your answer reasonable?

DYNAMICS WORKSHEET Name _____ Problem _____

PREPARE

- List knowns. Identify what you're trying to find.
- Identify forces and draw a free-body diagram.

- Draw a pictorial representation for problems with motion: Show important points in the motion, establish a coordinate system, define symbols, draw a motion diagram.

Known

Find

SOLVE

Start with Newton's first or second law in component form, adding other information as needed to solve the problem.

ASSESS

Have you answered the question?

Do you have correct units, signs, and significant figures?

Is your answer reasonable?

DYNAMICS WORKSHEET Name _____ Problem _____

PREPARE
- List knowns. Identify what you're trying to find.
- Identify forces and draw a free-body diagram.

- Draw a pictorial representation for problems with motion: Show important points in the motion, establish a coordinate system, define symbols, draw a motion diagram.

Known

Find

SOLVE
Start with Newton's first or second law in component form, adding other information as needed to solve the problem.

ASSESS
Have you answered the question?
Do you have correct units, signs, and significant figures?
Is your answer reasonable?

DYNAMICS WORKSHEET Name _____ Problem _____

PREPARE

- List knowns. Identify what you're trying to find.
- Identify forces and draw a free-body diagram.

- Draw a pictorial representation for problems with motion: Show important points in the motion, establish a coordinate system, define symbols, draw a motion diagram.

Known

Find

SOLVE

Start with Newton's first or second law in component form, adding other information as needed to solve the problem.

ASSESS

Have you answered the question?
Do you have correct units, signs, and significant figures?
Is your answer reasonable?

DYNAMICS WORKSHEET   Name _____   Problem _____

PREPARE

- List knowns. Identify what you're trying to find.
- Identify forces and draw a free-body diagram.

- Draw a pictorial representation for problems with motion: Show important points in the motion, establish a coordinate system, define symbols, draw a motion diagram.

Known

Find

SOLVE

Start with Newton's first or second law in component form, adding other information as needed to solve the problem.

ASSESS

Have you answered the question?

Do you have correct units, signs, and significant figures?

Is your answer reasonable?

DYNAMICS WORKSHEET   Name _____   Problem _____

PREPARE

- List knowns. Identify what you're trying to find.
- Identify forces and draw a free-body diagram.

- Draw a pictorial representation for problems with motion: Show important points in the motion, establish a coordinate system, define symbols, draw a motion diagram.

Known

Find

SOLVE

Start with Newton's first or second law in component form, adding other information as needed to solve the problem.

ASSESS

Have you answered the question?
Do you have correct units, signs, and significant figures?
Is your answer reasonable?

DYNAMICS WORKSHEET Name _____ Problem _____

PREPARE

- List knowns. Identify what you're trying to find.
- Identify forces and draw a free-body diagram.

- Draw a pictorial representation for problems with motion: Show important points in the motion, establish a coordinate system, define symbols, draw a motion diagram.

Known

Find

SOLVE

Start with Newton's first or second law in component form, adding other information as needed to solve the problem.

ASSESS

Have you answered the question?

Do you have correct units, signs, and significant figures?

Is your answer reasonable?

MOMENTUM WORKSHEET Name _____ Problem _____

PREPARE
- Identify the system and any external forces.
- Sketch "before and after."
- Define coordinates.
- List knowns. Identify what you're trying to find.

Known

Find

- Is momentum conserved? _____

SOLVE
Start with conservation of momentum or the impulse-momentum theorem, using Newton's laws or kinematics as needed.

ASSESS
Have you answered the question?
Do you have correct units, signs, and significant figures?
Is your answer reasonable?

MOMENTUM WORKSHEET Name _____ Problem _____

PREPARE
- Identify the system and any external forces.
- Sketch "before and after."
- Define coordinates.
- List knowns. Identify what you're trying to find.

Known

Find

- Is momentum conserved? _____

SOLVE

Start with conservation of momentum or the impulse-momentum theorem, using Newton's laws or kinematics as needed.

ASSESS
Have you answered the question?
Do you have correct units, signs, and significant figures?
Is your answer reasonable?

MOMENTUM WORKSHEET Name _____ Problem _____

PREPARE
- Identify the system and any external forces.
- Sketch "before and after."
- Define coordinates.
- List knowns. Identify what you're trying to find.

Known

Find

- Is momentum conserved? _____

SOLVE
Start with conservation of momentum or the impulse-momentum theorem, using Newton's laws or kinematics as needed.

ASSESS
Have you answered the question?
Do you have correct units, signs, and significant figures?
Is your answer reasonable?

MOMENTUM WORKSHEET   Name _____   Problem _____

PREPARE
- Identify the system and any external forces.
- Sketch "before and after."
- Define coordinates.
- List knowns. Identify what you're trying to find.

Known

Find

- Is momentum conserved? _____

SOLVE
Start with conservation of momentum or the impulse-momentum theorem, using Newton's laws or kinematics as needed.

ASSESS
Have you answered the question?
Do you have correct units, signs, and significant figures?
Is your answer reasonable?

MOMENTUM WORKSHEET   Name _____   Problem _____

PREPARE

- Identify the system and any external forces.
- Sketch "before and after."
- Define coordinates.
- List knowns. Identify what you're trying to find.

Known

Find

- Is momentum conserved? _____

SOLVE

Start with conservation of momentum or the impulse-momentum theorem, using Newton's laws or kinematics as needed.

ASSESS

Have you answered the question?
Do you have correct units, signs, and significant figures?
Is your answer reasonable?

MOMENTUM WORKSHEET Name _____ Problem _____

PREPARE
- Identify the system and any external forces.
- Sketch "before and after."
- Define coordinates.
- List knowns. Identify what you're trying to find.

Known

Find

- Is momentum conserved? _____

SOLVE
Start with conservation of momentum or the impulse-momentum theorem, using Newton's laws or kinematics as needed.

ASSESS
Have you answered the question?
Do you have correct units, signs, and significant figures?
Is your answer reasonable?

MOMENTUM WORKSHEET Name _____ Problem _____

PREPARE

- Identify the system and any external forces.
- Sketch "before and after."
- Define coordinates.
- List knowns. Identify what you're trying to find.

Known

Find

- Is momentum conserved? _____

SOLVE

Start with conservation of momentum or the impulse-momentum theorem, using Newton's laws or kinematics as needed.

ASSESS

Have you answered the question?
Do you have correct units, signs, and significant figures?
Is your answer reasonable?

MOMENTUM WORKSHEET Name _____ Problem _____

- Identify the system and any external forces.
- Sketch "before and after."
- Define coordinates.
- List knowns. Identify what you're trying to find.

Known

Find

- Is momentum conserved? _____

SOLVE

Start with conservation of momentum or the impulse-momentum theorem, using Newton's laws or kinematics as needed.

ASSESS

Have you answered the question?

Do you have correct units, signs, and significant figures?

Is your answer reasonable?

MOMENTUM WORKSHEET Name _____ Problem _____

PREPARE
- Identify the system and any external forces.
- Sketch "before and after."
- Define coordinates.
- List knowns. Identify what you're trying to find.

Known

Find

- Is momentum conserved? _____

SOLVE
Start with conservation of momentum or the impulse-momentum theorem, using Newton's laws or kinematics as needed.

ASSESS
Have you answered the question?
Do you have correct units, signs, and significant figures?
Is your answer reasonable?

MOMENTUM WORKSHEET Name _____ Problem _____

PREPARE
- Identify the system and any external forces.
- Sketch "before and after."
- Define coordinates.
- List knowns. Identify what you're trying to find.

Known

Find

- Is momentum conserved? _____

SOLVE
Start with conservation of momentum or the impulse-momentum theorem, using Newton's laws or kinematics as needed.

ASSESS
Have you answered the question?
Do you have correct units, signs, and significant figures?
Is your answer reasonable?

MOMENTUM WORKSHEET Name _____ Problem _____

PREPARE
- Identify the system and any external forces.
- Sketch "before and after."
- Define coordinates.
- List knowns. Identify what you're trying to find.

Known

Find

- Is momentum conserved? _____

SOLVE

Start with conservation of momentum or the impulse-momentum theorem, using Newton's laws or kinematics as needed.

ASSESS

Have you answered the question?
Do you have correct units, signs, and significant figures?
Is your answer reasonable?

MOMENTUM WORKSHEET Name _____ Problem _____

PREPARE
- Identify the system and any external forces.
- Sketch "before and after."
- Define coordinates.
- List knowns. Identify what you're trying to find.

Known

Find

- Is momentum conserved? _____

SOLVE
Start with conservation of momentum or the impulse-momentum theorem, using Newton's laws or kinematics as needed.

ASSESS
Have you answered the question?
Do you have correct units, signs, and significant figures?
Is your answer reasonable?

ENERGY WORKSHEET

Name _____ Problem _____

PREPARE

- Identify the system.
- Sketch "before and after."
- Define coordinates.
- List knowns. Identify what you're trying to find.

Known

Find

- Is the system isolated? _____ Which energies change? _____

SOLVE

Start with conservation of energy, adding other information and techniques as needed to solve the problem.

ASSESS

Have you answered the question?
Do you have correct units, signs, and significant figures?
Is your answer reasonable?

ENERGY WORKSHEET   Name _____   Problem _____

PREPARE
- Identify the system.
- Sketch "before and after."
- Define coordinates.
- List knowns. Identify what you're trying to find.

Known

Find

- Is the system isolated? _____   Which energies change? _____

SOLVE

Start with conservation of energy, adding other information and techniques as needed to solve the problem.

ASSESS
Have you answered the question?
Do you have correct units, signs, and significant figures?
Is your answer reasonable?

ENERGY WORKSHEET Name _____ Problem _____

PREPARE
- Identify the system.
- Sketch "before and after."
- Define coordinates.
- List knowns. Identify what you're trying to find.

Known

Find

- Is the system isolated? _____ Which energies change? _____

SOLVE

Start with conservation of energy, adding other information and techniques as needed to solve the problem.

ASSESS
Have you answered the question?
Do you have correct units, signs, and significant figures?
Is your answer reasonable?

ENERGY WORKSHEET

Name _____ Problem _____

PREPARE

- Identify the system.
- Sketch "before and after."
- Define coordinates.
- List knowns. Identify what you're trying to find.

Known

Find

- Is the system isolated? _____ Which energies change? _____

SOLVE

Start with conservation of energy, adding other information and techniques as needed to solve the problem.

ASSESS

Have you answered the question?

Do you have correct units, signs, and significant figures?

Is your answer reasonable?

ENERGY WORKSHEET

Name _____ Problem _____

PREPARE

- Identify the system.
- Sketch "before and after."
- Define coordinates.
- List knowns. Identify what you're trying to find.

Known

Find

- Is the system isolated? _____ Which energies change? _____

SOLVE

Start with conservation of energy, adding other information and techniques as needed to solve the problem.

ASSESS

Have you answered the question?
Do you have correct units, signs, and significant figures?
Is your answer reasonable?

ENERGY WORKSHEET

Name _____ Problem _____

PREPARE
- Identify the system.
- Sketch "before and after."
- Define coordinates.
- List knowns. Identify what you're trying to find.

Known

Find

- Is the system isolated? _____ Which energies change? _____

SOLVE

Start with conservation of energy, adding other information and techniques as needed to solve the problem.

ASSESS

Have you answered the question?

Do you have correct units, signs, and significant figures?

Is your answer reasonable?

ENERGY WORKSHEET

Name _____ Problem _____

PREPARE

- Identify the system.
- Sketch "before and after."
- Define coordinates.
- List knowns. Identify what you're trying to find.

Known

Find

- Is the system isolated? _____ Which energies change? _____

SOLVE

Start with conservation of energy, adding other information and techniques as needed to solve the problem.

ASSESS

Have you answered the question?
Do you have correct units, signs, and significant figures?
Is your answer reasonable?

ENERGY WORKSHEET

Name _____ Problem _____

PREPARE

- Identify the system.
- Sketch "before and after."
- Define coordinates.
- List knowns. Identify what you're trying to find.

Known

Find

- Is the system isolated? _____ Which energies change? _____

SOLVE

Start with conservation of energy, adding other information and techniques as needed to solve the problem.

ASSESS

Have you answered the question?
Do you have correct units, signs, and significant figures?
Is your answer reasonable?

ENERGY WORKSHEET

Name _____   Problem _____

PREPARE

- Identify the system.
- Sketch "before and after."
- Define coordinates.
- List knowns. Identify what you're trying to find.

Known

Find

- Is the system isolated? _____   Which energies change? _____

SOLVE

Start with conservation of energy, adding other information and techniques as needed to solve the problem.

ASSESS

Have you answered the question?
Do you have correct units, signs, and significant figures?
Is your answer reasonable?

ENERGY WORKSHEET

Name _____ Problem _____

PREPARE
- Identify the system.
- Sketch "before and after."
- Define coordinates.
- List knowns. Identify what you're trying to find.

Known

Find

- Is the system isolated? _____ Which energies change? _____

SOLVE

Start with conservation of energy, adding other information and techniques as needed to solve the problem.

ASSESS
Have you answered the question?
Do you have correct units, signs, and significant figures?
Is your answer reasonable?

ENERGY WORKSHEET

Name _____ Problem _____

PREPARE
- Identify the system.
- Sketch "before and after."
- Define coordinates.
- List knowns. Identify what you're trying to find.

Known

Find

- Is the system isolated? _____ Which energies change? _____

SOLVE
Start with conservation of energy, adding other information and techniques as needed to solve the problem.

ASSESS
Have you answered the question?
Do you have correct units, signs, and significant figures?
Is your answer reasonable?

ENERGY WORKSHEET

Name _____ Problem _____

PREPARE

- Identify the system.
- Sketch "before and after."
- Define coordinates.
- List knowns. Identify what you're trying to find.

Known

Find

- Is the system isolated? _____ Which energies change? _____

SOLVE

Start with conservation of energy, adding other information and techniques as needed to solve the problem.

ASSESS

Have you answered the question?
Do you have correct units, signs, and significant figures?
Is your answer reasonable?

ENERGY WORKSHEET

Name _____ Problem _____

PREPARE
- Identify the system.
- Sketch "before and after."
- Define coordinates.
- List knowns. Identify what you're trying to find.

Known

Find

- Is the system isolated? _____ Which energies change? _____

SOLVE

Start with conservation of energy, adding other information and techniques as needed to solve the problem.

ASSESS

Have you answered the question?
Do you have correct units, signs, and significant figures?
Is your answer reasonable?

ENERGY WORKSHEET

Name _____ Problem _____

PREPARE

- Identify the system.
- Sketch "before and after."
- Define coordinates.
- List knowns. Identify what you're trying to find.

Known

Find

- Is the system isolated? _____ Which energies change? _____

SOLVE

Start with conservation of energy, adding other information and techniques as needed to solve the problem.

ASSESS

Have you answered the question?

Do you have correct units, signs, and significant figures?

Is your answer reasonable?

ENERGY WORKSHEET

Name _____   Problem _____

PREPARE
- Identify the system.
- Sketch "before and after."
- Define coordinates.
- List knowns. Identify what you're trying to find.

Known

Find

- Is the system isolated? _____   Which energies change? _____

SOLVE
Start with conservation of energy, adding other information and techniques as needed to solve the problem.

ASSESS
Have you answered the question?
Do you have correct units, signs, and significant figures?
Is your answer reasonable?

ENERGY WORKSHEET

Name _____ Problem _____

PREPARE

- Identify the system.
- Sketch "before and after."
- Define coordinates.
- List knowns. Identify what you're trying to find.

Known

Find

- Is the system isolated? _____ Which energies change? _____

SOLVE

Start with conservation of energy, adding other information and techniques as needed to solve the problem.

ASSESS

Have you answered the question?

Do you have correct units, signs, and significant figures?

Is your answer reasonable?

ENERGY WORKSHEET

Name _____ Problem _____

PREPARE

- Identify the system.
- Sketch "before and after."
- Define coordinates.
- List knowns. Identify what you're trying to find.

Known

Find

- Is the system isolated? _____ Which energies change? _____

SOLVE

Start with conservation of energy, adding other information and techniques as needed to solve the problem.

ASSESS

Have you answered the question?
Do you have correct units, signs, and significant figures?
Is your answer reasonable?

ENERGY WORKSHEET

Name _____ Problem _____

PREPARE
- Identify the system.
- Sketch "before and after."
- Define coordinates.
- List knowns. Identify what you're trying to find.

Known

Find

- Is the system isolated? _____ Which energies change? _____

SOLVE

Start with conservation of energy, adding other information and techniques as needed to solve the problem.

ASSESS

Have you answered the question?

Do you have correct units, signs, and significant figures?

Is your answer reasonable?

ENERGY WORKSHEET

Name _____ Problem _____

PREPARE

- Identify the system.
- Sketch "before and after."
- Define coordinates.
- List knowns. Identify what you're trying to find.

Known

Find

- Is the system isolated? _____ Which energies change? _____

SOLVE

Start with conservation of energy, adding other information and techniques as needed to solve the problem.

ASSESS

Have you answered the question?
Do you have correct units, signs, and significant figures?
Is your answer reasonable?

ENERGY WORKSHEET

Name _____   Problem _____

PREPARE

- Identify the system.
- Sketch "before and after."
- Define coordinates.
- List knowns. Identify what you're trying to find.

Known

Find

- Is the system isolated? _____   Which energies change? _____

SOLVE

Start with conservation of energy, adding other information and techniques as needed to solve the problem.

ASSESS

Have you answered the question?

Do you have correct units, signs, and significant figures?

Is your answer reasonable?

ENERGY WORKSHEET

Name _____ Problem _____

PREPARE
- Identify the system.
- Sketch "before and after."
- Define coordinates.
- List knowns. Identify what you're trying to find.

Known

Find

- Is the system isolated? _____ Which energies change? _____

SOLVE

Start with conservation of energy, adding other information and techniques as needed to solve the problem.

ASSESS

Have you answered the question?
Do you have correct units, signs, and significant figures?
Is your answer reasonable?

ENERGY WORKSHEET

Name _____ Problem _____

PREPARE

- Identify the system.
- Sketch "before and after."
- Define coordinates.
- List knowns. Identify what you're trying to find.

Known

Find

- Is the system isolated? _____ Which energies change? _____

SOLVE

Start with conservation of energy, adding other information and techniques as needed to solve the problem.

ASSESS

Have you answered the question?
Do you have correct units, signs, and significant figures?
Is your answer reasonable?

ENERGY WORKSHEET   Name _____   Problem _____

PREPARE
- Identify the system.
- Sketch "before and after."
- Define coordinates.
- List knowns. Identify what you're trying to find.

Known

Find

- Is the system isolated? _____   Which energies change? _____

SOLVE
Start with conservation of energy, adding other information and techniques as needed to solve the problem.

ASSESS
Have you answered the question?
Do you have correct units, signs, and significant figures?
Is your answer reasonable?

ENERGY WORKSHEET

Name _____ Problem _____

PREPARE

- Identify the system.
- Sketch "before and after."
- Define coordinates.
- List knowns. Identify what you're trying to find.

Known

Find

- Is the system isolated? _____ Which energies change? _____

SOLVE

Start with conservation of energy, adding other information and techniques as needed to solve the problem.

ASSESS

Have you answered the question?
Do you have correct units, signs, and significant figures?
Is your answer reasonable?